100 Years Exploring Life
1888–1988

The Marine Biological Laboratory at Woods Hole

100 Years Exploring Life, 1888–1988

The Marine Biological Laboratory at Woods Hole

Jane Maienschein
Department of Philosophy
Arizona State University

Selection and arrangement
of photographs by

Ruth Davis
Archivist, Marine Biological Laboratory

Jones and Bartlett Publishers
BOSTON

Editorial, Sales, and Customer Service Offices
Jones and Bartlett Publishers
20 Park Plaza
Boston, MA 02116

Printed in the United States of America
10 9 8 7 6 5 4 3 2 1

Library of Congress Cataloging-in-Publication Data

Maienschein, Jane.
100 years exploring life, 1888–1988 : the Marine Biological Laboratory at Woods Hole/ Jane Maienschein with selection and arrangement of photographs by Ruth Davis.
p. cm.
Bibliography: p.
ISBN 0–86720–120–7
1. Marine biological laboratory (Woods Hole, Mass.)—History.
I. Title.
QH91.M237 1989 88–8529
574′.072074492—dc19 CIP

Printing of text by Lancaster Press, Inc.,
Lancaster, PA.

Contents

Foreword

IMAGINE A WARM SUMMER NIGHT in mid-August. You are standing on a dock at midnight in the light of a full moon, peering down into the dark water below. The surface of the water ripples as if it were alive. As your eyes grow accustomed to the moonlit darkness, you realize that the water *is* alive. You are watching the mating ritual of thousands of writhing, swirling polychete worms as they seek out their mates under the influence of the full moon's light. A timeless event unfolds before you, connecting you back to primeval oceans, where ancient progenitors of these polychetes repeated a similar ritual, before any human eyes were there to see. Before you is an expression of life at its most fundamental, its most dynamic, its most breathtaking. Like thousands of students of marine embryology who stood on this dock before you, you will always remember this scene and the institution that led you to it: the Marine Biological Laboratory in Woods Hole, Massachusetts.

In 1988 the Marine Biological Laboratory (MBL) celebrated its centennial. As it enters its second century, the MBL not only looks back to a great past, but forward to a great future as the United States' premier biological research institute. It is therefore fitting that its unique story be told at this time. In this well-researched, sometimes humorous, always human "biography" of this eclectic institution, historian of science Jane Maienschein has caught a glimpse of what it is that has made the MBL so special to all who have spent any time there.

Essayist Lewis Thomas has described the MBL as a "National Biological Laboratory," an institution that brings together each summer a collection of biologists from across the United States and abroad. From its founding in 1888 onward, the MBL has indeed served as a gathering spot for biologists who come to Woods Hole not only to work with their favorite marine organisms, but also to converse with each other and exchange ideas in a way that seldom happens in the more limited confines of university biology departments.

Almost from the beginning, the MBL attracted international as well as American investigators, becoming increasingly a "World Biological Laboratory" since the turn of the century. Unlike its European counterparts, where the focus has always been almost exclusively on research, often by only the most established senior investigators, the MBL has always had a diversity of programs and personnel. There is, of course, a major emphasis on research. In addition, however, the MBL has always been equally devoted to teaching, running a number of summer courses for graduate students or others seeking to learn about a field or refresh that knowledge. Nobel laureates and high school teachers have been involved in the same MBL course during a given summer.

Through its research and its courses, the MBL has served as a nerve center for the development and propagation of biological knowledge in the twentieth century. For this distinction it owes a debt to the great European biological stations established a generation earlier: Roscoff and the Stazione Zoologica in Naples (1872), Villefranche-sur-mer, and Plymouth, England (1888), among others. It also has roots in two earlier American precedents: the Annisquam Laboratory of the Woman's Education Association of the Boston Society of Naturalists (established 1879), and Louis Agassiz's Anderson School of Natural History on Penikese Island (established 1872). Like every great centenarian, the MBL has drawn upon this heritage to create its own unique personality. The biographical history of that personality is the subject of Professor Maienschein's volume.

While everyone who has spent any time at the MBL agrees it is unique, it is not so easy to capture and characterize that uniqueness for others. Some have emphasized the intellectual atmosphere, the constant interest and attention to science, and the thirty-five or more Nobel laureates who have at one time or another been directly associated with the institution. Others have emphasized the MBL's playful and relaxing aspect, referring to it as the "summer camp for biologists." Still others have emphasized its social and sociological side, the fact that the MBL has nurtured the creative development of countless biologists in this country and abroad over the century, and that many people's affiliations later in life came from contacts they made with colleagues at the MBL as students or young investigators. (Many biologists even met their spouses at the MBL, either in courses or through research work.)

But, of course, none of these qualities by itself fully captures the MBL's uniqueness, since the MBL is, in a sense, a combination of all of them. It is a place where lots of science gets done—often very good science. In the early 1900s, for example, T. H. Morgan brought his fruit flies from New York to Woods Hole every summer just because the atmosphere for doing research and for exchanging ideas was so exciting. More recently, neurobiologists from NIH, Harvard, Columbia, and the University of California have used marine invertebrates, such as slugs, to study the neurological basis of behavior. The MBL is, indeed, a center for much of what is at the cutting edge of biological research today. At the same time, it is very much a place where people combine play and work, where discussions about repeated sequences of DNA, or neurotransmitter mechanisms, are punctuated by swimming, tennis, or boating. Yet the conversations are always resumed, often with fresh insights brought about by periods of relaxation and play. It is also a place where careers are made, and where new associations, professional and personal, are formed. Many collaborations have resulted from joint research work begun, or largely carried out, at the MBL. The uniqueness of the MBL lies in its *combination* of so many vital qualities.

I recently discussed the subject of the MBL's uniqueness with my colleague at Washington University, Viktor Hamburger, student of Nobel laureate Hans Spemann in the 1920s and himself an eminent embryologist, who spent a number of years as instructor, and later director, of the embryology course at the MBL (1936–46). I asked Professor Hamburger what he thought were the unique features of the institution. He replied that, for him, the uniqueness lies in the pastoral setting that allows people to have ongoing conversations with colleagues undistracted by the constant interruptions of daily life in the university. In addition, Hamburger noted, the MBL provides scientists the opportunity to observe each other's experiments first-hand, and to discuss different interpretations of data on the spot. The MBL provides an opportunity for doing science and for thinking in a relaxing atmosphere that fosters creativity. To Hamburger, this is the secret of the MBL's uniqueness—that it spawns creativity.

My own experience with the MBL dates back some twenty years when, as a graduate student in history of science, I discovered one of the Laboratory's most unique features: its magnificent library. Even though I was a student at Harvard, which has one of the most complete libraries in the world, I still found the MBL Library to be a special treat. Not only were the library's journal holdings more complete in many cases than Harvard's, but they were much easier to use. All numbers of every journal are housed under one roof and are arranged alphabetically. The library is open twenty-four hours a day, seven days a week, and no one checks you in and out or stamps the books you borrow; the library runs on the honor system and works surprisingly well. Few books are lost, stolen, or misplaced. Another very special feature of the MBL Library is its reprint collection: 300,000 individual reprints, arranged by author and covering all aspects of biology between roughly 1880 and 1966. For a historian, this is a goldmine of information.

After spending my first summer using the library full-time, I discovered—in a way close to my own heart—one of the MBL Library's most treasured features. Here was a complete set of the *Journal of the Royal Microscopical Society*, or the *Proceedings of the Royal Society of London*, dating back to volume 1, in 1665, sitting on the shelves for browsing. (Some of the older, more valuable and fragile items have now been removed to the newly-renovated Rare Books Room and Archives.) These same journals had been perused in the past by Lillie, Morgan, Conklin, Harrison and other greats in early twentieth-century biology.

In the two decades since, I have never failed to feel the excitement of this library each day—as much, perhaps, as the embryologist still feels each time he or she looks at a developing ctenophore or sea urchin embryo. I am reminded of the tribute Stephen Jay Gould paid to the MBL Library in his book, *Ontogeny and Phylogeny*, written partly in Woods Hole over a decade ago: "Where else could an idiosyncratic worker like me find a library open all the time, free from the rules and bureaucracy that stifle scholarship and 'protect' books only by guarding them from use. It is an anomaly in a suspicious and anonymous age."

It is difficult to capture these many and varied facets of the MBL's personality in one book, especially one aimed at a general audience of nonspecialists. If you have spent one or more summers at the MBL, you know some of the magic that it holds. But to someone who only knows *of* the MBL, or is just learning about it, the qualities that make it so special to its friends may seem elusive, almost mystical.

It is a testament of Jane Maienschein's knowledge of the MBL and to her historical and verbal skills that she has captured much of the magic of the institution in a very down-to-earth way. She has truly written a *biography* of the laboratory. Professor Maienschein has chosen to write not so much of the Nobel laureates or of the detailed scientific accomplishments that have made the MBL's first century so eminent, but of the people—the everyday people, scientists and nonscientists alike—who have made up the life of the Laboratory and created its very special personality.

This is the story of an institution written in highly personal terms. The Nobel laureates are there, as they should be. So, too, are the directors, trustees and others who have given special parts of their lives to guiding and managing the institution. But it is the working scientists and support staff—the collecting crews, the technicians, even the doyens of the Mess Hall—whose story Professor Maienschein paints with such clarity and humor. The work-a-day activities of people from every facet of MBL life occupy the main focus of Professor Maienschein's attentions. It is, after all, the activity of everyday people that makes up the real spirit of any institution. That spirit existed from the beginning at the MBL, before there were Nobel prizes or year-round administrative staff. In many ways, it would have been easier to write a book that traced only the illustrious individuals who figured prominently in the MBL during its first century. The more difficult task, which Professor Maienschein has carried out so well, is to portray that history in terms of on-going, everyday activities.

Although Jane Maienschein has served as principal author, in a very real sense this book is a collaborative effort. Some of the collaborators are long deceased, but their voices are heard in the quotations from archival sources, both published and unpublished, that Professor Maienschein has included. Others of the collaborators are still very much with us, and they have collaborated by providing interviews and personal recollections that always add so much to historical accounts.

Certain collaborators have added their own very special touches. In particular, MBL archivist Ruth Davis, along with Robert and Millie Huettner, has been invaluable in selecting the photographs to illustrate the volume. This history of the MBL, therefore, is the work of a team of researchers, and presents the best of their collective effort.

As a result of this collaborative approach, Professor Maienschein's book is unique in its own right. It is intended as a living testament to an institution with a special mission and a special history. It is not a sequel to or replacement of F. R. Lillie's book, *The Woods Hole Marine Biological Laboratory* (Chicago, University of Chicago Press, 1944), which has been reprinted in a special edition for the MBL centennial by Lancaster Press. Nor is Professor Maienschein's book an official centennial eulogy, an essay in self-aggrandizement of the sort that often accompanies the celebration of institutional anniversaries. Rather, it is what any good biography should be: a loving but frank portrayal of a special friend.

Like all celebrities, the MBL has had its ups and downs. The ups have far outnumbered the downs, but the downs have been there nonetheless: the perennial financial problems, disagreements within the corporation about its own mission, and even controversial proposals to have the MBL managed by other institutions. But the MBL has managed to come through these bad times more or less unscathed, retaining, as Jane Maienschein shows so well, its essential magic.

This book is aimed at the specialist and the nonspecialist alike. It is written so as not to presuppose any particular background, or any familiarity with biology or the MBL itself. As a personal history, it should be accessible not only to biologists who know the MBL first-hand, but also to the curious reader, the Cape Cod visitor, or the foreign dignitary who wants to know something about American scientific research institutions. It may, therefore, disappoint some research biologists, who might wish that there were more detailed descriptions of the scientific work of the past or present at the MBL. It may also disappoint some staunch MBL regulars, who would like to see more details of local history and institutional lore. It may even disappoint that special breed of modern historian of science, the institutional historian, who prizes quantitative data—graphs and tables of numbers of investigators per summer, numbers of dollars spent, square footage of lab space utilized. All that would be largely irrelevant, however. It would not reveal nearly so well as does Professor Maienschein's more personal account what it is that really makes up the spirit of the MBL. No book can be all things to all people. One must take this portrait of the MBL on its own terms.

A special feature of Professor Maienschein's book is the many and varied illustrations that accompany it. This book is really more of an "illustrated biography" than a standard history. Included are many unusual photographs and prints taken largely from the MBL Archives and selected with great care by archivist Ruth Davis.

In line with Professor Maienschein's overall approach, most of the photographs are candid, rather than posed, portraits. The illustrations attempt to capture the liveliness, the spirit of intense work, humor, play, and family that make up so much of life in the MBL community. Selected from all periods of the MBL's history, the illustrations attempt to show, in contemporary terms, the kind of scientific work, physical atmosphere, and personal relationships that existed in 1888, 1910, 1940, 1980. If a picture is worth a thousand words, Professor Maienschein has lengthened her text considerably by including this array of fascinating and informative illustrations.

In conclusion, this institutional biography is written from the perspective that could only have been written by a professional historian who also has an extended and loving relationship with the MBL. It is as fitting a way as I can imagine both to look back over the MBL's distinguished past and to glance forward to its equally exciting and promising future.

Fall 1988

GARLAND E. ALLEN

Preface

THE MARINE BIOLOGICAL LABORATORY began in 1888. As first director Charles Otis Whitman said, it was a mere germ, only barely fertilized. The first year brought a simple cellular form with only seventeen "ids in its protoplasmic body—two instructors, eight students, and seven investigators (all beginners). The two instructors could be likened, with no great stretch of the imagination, to two polar corpuscles, signifying little more than that the germ was a fertile one, and prepared to begin its preordained course of development." This fertile germ then underwent various cleavages and began to assume a multicellular shape. With growth, it advanced to the tadpole stage. It encountered troubles along the way, just as any growing individual does. Fortunately, these troubles have never proved fatal.

Whitman saw the MBL as a living being. It still is today—a being made up of all the cellular individuals who visit and work there. The life of the organism is part of the life of the individual members and visitors. This book is about the MBL's first one hundred years of life.

This work represents a biography of the MBL, which has had a life very like any other individual, with its cycles of adolescence and growth and maturity and maybe even metamorphic stages as well. Indeed, this is really an autobiography, a reflective sketch of a full and rich life told through recollections and reflections recorded within the archives and the people of the MBL. This is only one of the many autobiographical stories that could be told; other times and other efforts will bring additional perspectives. If anyone feels that something has been left out, he or she is invited to write down those stories and facts and deposit them in the MBL Archives. This story reflects what lies in the current public and archival record.

As an autobiography, this is not the sort of study that provides a litany of vital statistics and details from birth to death in precise, chronological order with everything in its proper place. Rather, this is an effort to present the spirit of the MBL's life, that spirit of cooperation and cross-fertilization of ideas that makes science a living process. This interactive life is sometimes pervaded by a certain untidiness and even unreality. Recorded memory occasionally may bring seemingly insignificant details into sharp focus or forget others. Sometimes the past comes into clearer focus than the present. So be it. The life recounted here is an important one, and the quirks of the storytelling mirror are quirks in the life itself and in its records.

What, exactly, is the MBL? Biologist-writer Lewis Thomas has called it a virtual National Biological Laboratory for the United States because so much of import has been accomplished here, by so many leading biologists from all over the country. Yet the MBL is not officially that. It is not funded by the government, nor does it have direct governmental connections. It is an independent research and teaching laboratory, owned and governed by its scientists, and it has been so since the early years. The MBL welcomes biologists from all over the country and from many foreign countries; one recent year brought representatives from over 325 American and nearly 75 foreign institutions.

This rich diversity of scientists clearly goes beyond any self-conscious sense of unique national identity or externally imposed purpose. Rather, the MBL is a group of first-rate individual scientists, working in concert with their own goals. These goals converge on producing the best biological research possible, so that the underlying purpose is one of advancing science. The MBL is a haven for science and as such serves as a special resource, nationally and internationally. It is more a valuable treasure than a national laboratory in the most familiar and limited sense.

Yet what, exactly, is the Marine Biological Laboratory, and where did it come from? It is mostly Marine. The majority of researchers still use marine organisms. Some of these are not strictly local; a few could easily be flown back to Idaho or Kansas and need not be studied on the spot; and a small number of workers do not really even work with marine organisms. The MBL has embraced a wider variety of life forms as the evolution of biological work has carried researchers elsewhere. Yet the bulk of work at the MBL remains marine and directed at marine-based research problems concerning development, heredity, physiology, and evolution.

The work is also largely Biological, though not exclusively. Chemistry and physics creep in as they relate to biological questions. Historians have begun to join the group of researchers, carrying out their own historical research projects using the unique collection of resources available. After the recent inaugural history course in connection with the MBL's centennial, several of the students cancelled their vacation plans in order to stay and do research in this surprisingly outstanding library. Occasional artists arrive, and journalists, and sociologists, to carry out their own brand of research, not in biology but about biology and biologists. Yet Biology remains the central mission.

As a Laboratory, the place also has evolved to embrace a wider range of work centered on biological research. Some of the researchers at the MBL actually use the library as their laboratory, a few even making the mistake of never going outside to explore the variety of life there. Some take the institution as their lab, looking over the shoulders of scientists to investigate the process of doing science. But most still come to the MBL to carry out their laboratory work in biology, experimenting on marine organisms as they have for a full century. In recent decades the MBL has become a year-round laboratory as well, with some outstanding researchers choosing to pursue their life's work here. One recent wintry day, someone put warm woolen hats on artist Elaine Pear Cohen's fine sculpture of scientists talking, which represents the MBL spirit and resides at a major corner in town. It evidently appeared that the scientists needed additional warming resources to continue through winter.

At first the MBL was just a summer operation. It began in 1888 as part of America's response to the general move toward research at the seashore. Little was known of marine life before 1800, but that began to change from two directions. The laying of transoceanic telegraph cable brought many questions about marine life. Common opinion had held that the pressure of water would prevent any life from existing at very great depths in the seas. People expected to find neat layers of different sorts of living beings, below which would lie a layer of skeletons from those human bodies lost and buried at sea, and below that perhaps a layer of gold coins, and anchors, and other items lost overboard. Yet when the deep sea cables broke and were hauled up for repair, they had numerous living organisms securely attached. Life forms must be able to live down there after all, and the drive quickly developed to explore those depths and to discover those living organisms "where no man had gone before." Perhaps the sea even held very simple and primitive organisms that would help to illuminate the perfection of nature's design, thought researchers in the middle of the nineteenth century, before Darwin.

In the late nineteenth century, after Darwin had put forth his evolution theory and after German biologist Ernst Haeckel had convinced so many people that *the* right way to pursue biological science was to trace the evolutionary relationships of organisms, biologists moved to the seashore. Haeckel believed that all of life arose from very simple primitive organisms resembling single cells. The question was, which organisms appeared first and which later, and through what series of changes? The key, Haeckel convinced a number of researchers, lay in marine organisms. Sea life, he believed, was more primitive and therefore more basic in evolutionary history. Studying the similarities and differences, especially of early embryonic development of a range of marine organisms, became the accepted practice in biology. Besides, knowing the phylogenetic history, as it was called, would reveal the ancestors of the vertebrates and of man. This, after all, is something we care about. So to the seashore they went.

Those few hardy pioneers therefore moved to the seashore to inves tigate the structure and function of the various peculiar aquatic species. They asked: what lived in the water, and how did those forms relate to terrestrial organisms? What could marine life reveal about the marvelous diversity and distribution of life? The European approach to the sea centered on facilitating research into just such questions. In contrast to the dominant research-oriented Naples Zoological Station in Italy and other European labs, the Americans sought to establish a teaching as well as a research laboratory, a place where landlocked and uninitiated students could experience scientific investigation with living organisms at the seashore. As the founders wed their two goals together the MBL emerged. That MBL has trained a hundred years of biologists, directly and indirectly. Each summer hundreds of students come from all over to take courses taught by teams of hundreds of internationally known lecturers (yes, there really are at least as many lecturers as students). One-time students often go on to become lecturers in their turn, or they help to start and sustain other marine laboratories and schools elsewhere, many of which have carried on some of the MBL tradition.

So the MBL is just that: the Marine Biological Laboratory, with a life of its own and an identity that defies neat and tidy circumscription. It is a place where people learn to love being part of the process of doing science, and where the science benefits too. And it is an exemplar for community research in biology, a hotbed of intense, dedicated biological research. It is a place where one can look out into an audience gathered for a lecture and see two new MacArthur Fellows sitting next to each other, where National Academy of Sciences members abound, where there are several NIH Merit researchers together, where department heads and Nobel Prize winners congregate. More importantly, it is a place where those recognized by such external distinctions sit and discuss the same lectures and the same research data or course outlines as young assistant professors, eager graduate and undergraduate students, enthusiastic high school students, and sometimes even the children. For this is a place to learn the sharing and cooperation that makes cross-fertilization of ideas possible, to ignore or overcome the boundaries existing elsewhere in the research world. It is a place to return to and to work for and to care for. This book is a story of the MBL, told through the words and images of its people, both past and present.

J.M.

NOTES*

Charles Otis Whitman on the beginning of the MBL, "Address to the MBL Corporation," August 11, 1903, Whitman Collection.

Lewis Thomas on the MBL as a national laboratory, *The Lives of a Cell* (New York: Viking Press, 1974), pp. 58–63.[0]

On early marine work, see the symposium papers by Keith Benson, Ralph Dexter, Jane Maienschein, and Robert Terwilliger, "The History of Marine Laboratories and Marine Research," *American Zoologist* (1988) 28: 1–34.

On the Naples Zoological Station, see Charles Kofoid, *The Biological Stations of Europe* (Washington, D. C.: Government Printing Office, 1910), pp. 7–34. Also more recent articles by Christiane Groeben, such as "Anton Dohrn—the Statesman of Darwinism," *Biological Bulletin* special historical edition (June 1985), 168 Suppl.: 4–25.

*Materials are in the MBL Archives unless otherwise noted.

1
Arriving in Woods Hole

The Yalden sundial, given in 1934 by Charles R. Crane. MBL Archives.

The Scientists, *sculpture by Elaine Pear Cohen, at the corner of School and Water streets.* Photograph by Sally Bruckner, courtesy of Elaine Pear Cohen, MBL Archives.

James P. McGinnis injecting a lobster, 1958. MBL Archives.

FRIDAY NIGHT in the summer, the cars roll into Woods Hole. Eager families arrive for their ferry reservations to Martha's Vineyard. They have waited a long time for this week to come, and they arrive early in anticipation. Steamship Authority officials direct their cars into line. But hours remain before the scheduled departure. The family locks the car and wanders the streets of Woods Hole, this little piece of land at the bottom of Cape Cod. The main street takes them past restaurants, specializing in seafood of course, then past the grocery store. But what are the red brick buildings? Is there some business or factory here, they ask? Only the business of science.

As they move down the street, they may see a sundial on a little plaza near the water. An MBL benefactor, plumbing magnate Charles R. Crane, gave the Yalden sundial, which was designed to keep extremely accurate time—reportedly to within one-half minute throughout the day—for Woods Hole's particular location. A lobster provides the sundial with its local Cape Cod flavor and a scientific twist. There is a small hole in the lobster, it is said, because the stone engraver obtained his specimen from the MBL Supply Department rather than from a local fisherman. Reportedly Harvard biologist George Howard Parker initially told the engraver that the claws were not lifelike enough. He recommended a visit to the supply department. The chosen model had a little hole in the back of the carapace, where it had been injected. So does the sundial's lobster.

Friday Evening Lectures

Very few of those waiting for the ferry make the turn past the lobster down MBL Street, to the brick and wooden buildings and to the Eel Pond. Those few who do so on a Friday night will see people converging from all directions on the main lecture room of the MBL, in the Lillie Building. Individuals trickle in from the library floors above, some with books in hand. Groups of people wander over from the laboratories across the street or down the hall, often chattering away about what is working or not

Crane Building from the old Cayadetta *dock, August, 1923.*
Norman W. Edmund Collection, MBL Archives.

Lillie Auditorium during the 1940s, with director Charles Packard sitting in the center background. MBL Archives.

Charles Otis Whitman, first director of the MBL and architect of its form of governance of, by, and for the scientists. MBL Archives.

working in the lab. Some people drive in from their cottages, often transporting those whose age has made them feel no longer comfortable walking the familiar distance at night. Bicycles roll up carrying young researchers, students, and maybe even a few teenagers. Assorted people sail or row in after a day on the water. Maybe a couple will even be returning by ferry from a day on one of the islands.

These people are all gathering, as the MBL community has gathered for one hundred years, to hear the week's Friday evening lecture. Some nights the subject may be so technical that a few tired listeners doze off after their long day in the lab. At other times a brilliant lecturer will enchant everyone with carefully chosen examples, beautiful slides, and a persuasive explanation of why this work matters. This latter type of lecture is what the Friday evening lectures have always been about.

From the very first years, director Charles Otis Whitman felt that even a specialized modern laboratory such as the MBL should have a time when the entire community would come together to consider the major scientific problems of the day. Individuals should learn from each other, he urged. People should cooperate even as they pursue their separate research. Regular lectures to address the key problems, as well as to discuss the most effective methods of approach and the best available explanations, should also be able to illustrate to the public what marine biological research is about.

With time those lectures have become somewhat more specialized and more technical, but always with emphasis on presenting the latest concerns of the day. Recent years have spawned additional series of general lectures, by journalists on one day and by historians and philosophers on another. Discussing science forms an essential part of MBL life, so the visitor should not be surprised to find all those people moving willingly inside to sit in a lecture hall and listen, even on a perfectly gorgeous and inviting summery afternoon or Friday evening.

In fact, the tradition of lecturing to the public as well as the scientific community about biological topics has been a major part of American culture and science for a long time. Boston in the nineteenth century had its naturalist Louis Agassiz, who may well have initially helped to inspire the MBL's Friday evening series.

Louis Agassiz

Agassiz came to this country from Switzerland in 1846 to give a lecture tour. He loved to travel and to explore the world and had somehow heard about the delights of lecturing in America. No one worried that his English was not perfect. Science was gaining popularity, and the word that he was a fine speaker was enough for an agent to book a tour for him. In Boston Agassiz

Louis Agassiz at the blackboard.
MBL Archives.

spoke to a crowded audience of 1,200 people. When he occasionally had to pause and grasp for the proper English word, he also grasped the chalk and began to draw. Sometimes using both hands, Agassiz drew organisms that came alive on the board behind him, entertaining the audience further.

Naturalists in the nineteenth century had to draw, because they often spent a high percentage of their time meticulously depicting what they had seen in order to communicate it to others who had not. Naturalists who could not draw had serious trouble. Or those who found painstaking watercolor work carelessly washed away by drops of seawater falling from wet hair after a swim could lose a precious investment. They could not simply photograph what they saw, not until the very end of the century, and then only with poorer clarity than the eye could see. They did not have the remarkable advanced technology, such as video microscopy, being developed today. The photographer that the MBL first added to the staff in 1897 did not replace practical everyday drawing for quite some time.

A tremendous popular success, Louis Agassiz liked the United States and determined, after the death of his first wife, to settle in Boston. There he married Elizabeth Cabot Cary, later president of Radcliffe College. In 1847 he became established at Harvard University, where he founded the Museum of Comparative Zoology. Always an opponent of evolution, his students once suggested to him that a debate between evolutionists and nonevolutionists might prove illuminating, all in the spirit of open scientific discussion and the search for truth, of course. Agassiz reportedly responded "rather evasively" that "personally I like Mr. Darwin very much; he is my friend." Someone then pointed out that "Darwin's son Frank [Francis] was once told that Agassiz did not accept evolution. 'That's all right,' said Frank, 'father does not believe in the glacial theory.' " Agassiz was very proud of his glacial theory, which held that much of the geological change that the earth has experienced has resulted from the cycling of glacial epochs. Apparently Agassiz chose not to pursue the discussion further at that point.

Three sketches from Frank Leslie's journal of August 23, 1873, showing scenes from the Anderson School of Natural History at Penikese: Louis Agassiz at the blackboard with chalk in hand, gentlemen dissecting a fish, and a room in the ladies' dormitory. Drawings by Albert Berghaus, MBL Archives.

In 1873 Boston, like much of America, was in the throes of popular enthusiasm for science when Agassiz arrived. The publicity in the second half of the nineteenth century for the notorious race for dinosaur bones by Yale's Othniel Marsh and Pennsylvania's Edward Drinker Cope had intensified public awareness of evolution theory and zoology generally. Nature study had gained great popularity here and abroad. Furthermore, the public wanted education in science, for themselves and for their children. Boston school committees mandated that there should be more science teaching in the schools, because there was little done officially, especially in biological subjects.

But who was to teach science? Who was to teach biology? Because biology at the time meant primarily natural history, and because natural history of organisms cannot be learned simply by peering at textbooks, any attempt to introduce biology into the schools meant having to teach the teachers. "Study nature, not books," Agassiz preached, explaining that he would not allow textbooks into his classroom. But how were these school teachers supposed to study nature?

In 1873 a student of Agassiz, Nathaniel Shaler, suggested that Agassiz offer a summer course for teachers. Yet the great popularizer had too many projects demanding his time and was no longer a young man. He was not sure. Besides it would take money to set up a school. He had enough trouble getting sufficient funds to keep his own Museum of Comparative Zoology (MCZ) at Harvard going in proper style. And where would such a school be? Giving a public lecture to generate support, he began to wonder out loud about the possibilities. As he appealed to the Massachusetts legislature for funds for the MCZ and for the summer school, it began to look as though a school might be possible, on Nantucket Island.

The Anderson School at Penikese

Then some of the ever-popular Louis Agassiz's concerns found their way into the New York newspapers, including the *Times* and *Tribune*. In response, a wealthy New York businessman, John Anderson, wrote to Agassiz offering land on the island of Penikese, off the coast of Woods Hole, plus his own house there for Agassiz's personal use: a gift valued at $100,000. He also gave $50,000 to serve as the base of a permanent endowment to open a summer school of natural history for teachers.

As reported in *Nature*, Anderson wrote to Agassiz that the school "may be destined in future ages not only to afford the required instruction to the youth of our country, but may be the means of attracting to our shores numerous candidates from the Old World, who may find here, in the school to be established by you, those means of fitting themselves for the teaching of Natural History by Nature itself. Which by a strange oversight, appears to

The Anderson School of Natural History on Penikese Island. The large building (center left) housed laboratories on the first floor and dormitories on the second. It burned to the ground in a spectacular fire in 1891. MBL Archives.

have been overlooked in the schemes . . . of education here." Agassiz and Anderson agreed to hold summer sessions on Anderson's Penikese property and winter sessions in Cambridge, in effect making the school the educational branch of the MCZ. After Agassiz had accepted and the final official papers had been signed in a fancy ceremony in New York, Anderson declared the event as "the happiest moment of my life."

The Anderson School opened in 1873 as planned, even though the construction of the main teaching building and the dorms continued up until the very last minute. Embarrassingly, too many people wanted to attend, perhaps in part because the price was so reasonable, with no tuition and only a percentage of the value of bedroom furniture to pay for room, plus board at cost. At first Agassiz had expected few applications and had simply accepted everyone who applied. But then some of the later applicants looked much more promising than the earlier ones. So Agassiz wrote a remarkable letter asking the earlier ones to surrender their places in favor of those later and better-prepared students. Evidently, in the face of such an unorthodox appeal, some did. The final select group of some forty students included a number of women who were "very schoolma'my in appearance," and the "gentlemen were not one whit behind," the press reported. The press attended the school's opening in force, for Agassiz recognized a good show and invited them in. Students, relatives, and members of the press all gathered in New Bedford* to take a special steamer to Penikese for a day. Agassiz was giving a scientific party.

At the laboratory of the new school, he presented a dedicatory address, which was widely acclaimed as "inspiring" and "beautiful" and as a "silent prayer," immortalized in John Greenleaf Whittier's often-cited ode "The Prayer of Agassiz," which reflects the ideals of the time:

*Unless otherwise indicated, all places mentioned are situated in Massachusetts.

On the Isle of Penikese,
Ringed about by sapphire seas,
Fanned by breezes salt and cool,
Stood the Master with his school. . . .
Said the Master to the youth:
'We have come in search of truth,
Trying with uncertain key
Door by door of mystery;
We are reaching, through His laws,
To the garment-hem of Cause . . .'

After the dedication the group enjoyed a feast. Then the reporters left the island to file their enthusiastic stories praising this great moment in American education and in American science. Only then, as one student put it, did the small group of remaining teachers-turned-students realize that they were settled on a barren island only two-thirds of a mile long and one-third of a mile wide. Without distractions, science would occupy the summer.

That summer's experience on that tiny island made a major difference to American science. Among those inspired by the island school was Alpheus Hyatt, destined to run his own teachers' school for the Boston Society of Natural History and to provide the motivating force for the founding of the MBL. There was also David Starr Jordan, who became the first president of Stanford University and sought to build it into a western scientific empire. Also among the students was Charles Otis Whitman, who taught at Boston's English High School. During this summer of 1873, and the next when he returned to Penikese as an advanced student under Agassiz's son Alexander (who took over the school for one year after Louis suddenly died), Whitman decided to become a professional biologist. That meant he would have to go to Europe to pursue a doctoral degree, with all the resulting difficulties for one from a "sober and pious" but definitely not wealthy family from Maine. But he had made up his mind, and Whitman was a stubborn and dedicated man. He went to Germany in 1875, where he received his Ph.D. for work on the embryology of leeches. After a two-year stay in Japan, he returned to the United States. He directed the Allis Lake Laboratory in Milwaukee, chaired the department at Clark University, and made his way to Woods Hole, full of ideas and energy, and ready to head the Marine Biological Laboratory when it began in 1888.

Summers in Woods Hole

Mt. Holyoke professor and former Penikese participant Cornelia Clapp arrived in the very first year of the MBL, in 1888, and found a quiet, tiny village. Ferry service from Woods Hole to Martha's Vineyard had begun in 1749, and hotels and summer estates had begun to appear by 1872, but only a few tourists and summer visitors had arrived yet by 1888. It was not until after the pungent guano works closed and was torn down after 1889 that Woods Hole really began to attract a significant population of tourists and others seeking refuge from the cities in seaside resorts.

In those days, one generally arrived by the train that first chugged into Woods Hole in 1872, though by mid-century one could have arrived after a pleasant overnight trip from New York by paddle-wheel steamer. Isabel Morgan Mountain, daughter of Columbia biologist Thomas Hunt Morgan, recalls that the family set out from New York for the summer in Woods Hole in great style. They took baskets of plants to grow in their Woods Hole garden, perhaps some mice to study, their English Setter, a pair of love birds, and the usual assortment of baggage and children. By train, whether the more local or the fancier "Dude" train, the trip took a little more than two hours from Boston, straight into Woods Hole along a track that brushes right up against the sand dunes and overlooks the water in a few places. When autos first bumped their way down the Cape, the trip was considerably longer; now it takes maybe one and a half hours, except on a Friday or Sunday in the summer tourist season. Now that the train is gone, bicycles, baby buggies, and pedestrians frequent the paved pathway where the tracks once lay.

Taking the train to Woods Hole made an impression on many newcomers, for there they often met other scientists and their families on their way to the summer haven. Second MBL director Frank Rattray Lillie reminisced in his unpublished autobiography that it was during the train trip that he first learned what the MBL was really about. Lillie had signed up to work as a graduate student at Clark University under Whitman. Whitman had suggested that he get himself to Woods Hole for the summer to begin work. As Lillie made his way from his home in Toronto to Woods Hole, he learned from fellow passengers that he would be expected to enter into the cell lineage research of the day. This work was dedicated to tracing, in a variety of organisms, what happens to each cell as it undergoes a series of divisions. The only real choice for a student, the others explained to Lillie, was which "beast" he would choose to study, and not which questions he would ask or which techniques he would employ. Those were set. For the sort of comparative enterprise that Whitman wanted, everyone had to standardize as many factors as possible.

By the time Lillie arrived in 1892, the MBL community all found housing in local boarding houses. The very first year, however, no one had made any housing arrangements for the students, and the choices were slim. Nor was there anywhere obvious to eat. In fact, when investigator Cornelia Clapp first arrived there really was no MBL. The carpenters were still working frantically on the one building and the director had not yet arrived. Eventually, the small group of students and instructors made arrangements for

Richard and Robert Huettner at the train station in Woods Hole in 1930. Photograph by Alfred F. Huettner, MBL Archives.

Although the date, location, and person in this picture are unknown, there is no question about where the freight was heading. F.R. Lillie papers, MBL Archives.

The Nantucket, *sidewheel ferry boat in Woods Hole Passage, with Devil's Foot Island in the foreground.* MBL Archives.

rooming and boarding at one of the few houses willing to take in strangers that first year. Or as Clapp reported, some took their meals in the "dark, dingy hole" of a dining room at the railroad station. They then picked their way through the glacial boulders toward the simple wooden building with its general open laboratory.

By 1902, in contrast, when student Beamon Douglas spent the summer at the MBL, it seemed to him that nearly every house in town eagerly took in boarders. He reported that "the modest sum of three or four dollars a week secures a large room, comfortable though simply furnished, with sufficient lamp oil, bed linens, and water for the most exacting." By then, the homeowners were often so glad of the opportunity to take in additional income for a few months that they rented out all available space to the busy and generally well-behaved scientists and themselves occupied the unrent-

William Procter in front of Old Main, 1923. Photograph by Alfred F. Huettner, MBL Archives.

The embryology class, 1894. Left to right: Henrietta L. Graves, Ellen Appleton Stone, Gilbert L. Houser, Wesley R. Coe, Julia Haynes; second row: Frances Crane (later Mrs. Frank R. Lillie), Charles S. Bacon, instructors Frank R. Lillie and Oliver S. Strong; Cresswell Shearer behind Lillie and Strong. Photograph by Baldwin Coolidge, MBL Archives.

able odd corners and closets. Douglas did not consider the prices unfair, but he regretted the crowdedness. Because the families took in so many roomers, Douglas felt that "association with them is not an unalloyed joy, and the student usually prefers to limit his acquaintance to the members of the laboratory mess." As one of the popular autograph books of the day suggested, "the Woods Hole landlord prefers his guests not to use his porch to entertain their callers"—another reason to stick close to the laboratory as the number of marriages among students and instructors at the MBL testifies to the interest in "calling."

A later story illustrates the sorts of concerns that the Woods Hole community long held about those scientists. When one biologist announced to his landlord that he intended to marry and return with his wife to live in the same home, the man asked what the new wife's name would be. When informed that she would take her husband's name, the landlord was reassured. "Oh," he said, "thank goodness that there are some people going to be here not living in sin."

Another feature of Woods Hole that has impressed visitors increasingly in the twentieth century is the lush greenery and flowers. Because the Cape's winters are moderated by the water, the area does not suffer much.

Old Main as it looked in 1892, set amidst the moraine which forms this part of Cape Cod.
MBL Archives.

The thick woods, protected by a series of agreements against wanton development in Woods Hole, provide a lovely place to take an evening's stroll—or to wish one could when the land is barricaded off as PRIVATE. Yet as Henry David Thoreau had pointed out with regret in his book *Cape Cod*, by the mid-nineteenth century the Cape had few forests any more. The sheep had eaten what people had not burned or cut (and the gypsy moths came along a few years later to carry out the last stages of deforestation). Thanks to the Boston merchant Joseph Fay, who sought to attract other summer residents to the area, the forests returned to Woods Hole. Fay planted an impressive twenty thousand mixed pine, larch, spruce, and birch seedlings to bring back the trees. Within fifty years, the forest around Woods Hole had reached a second growth or oak stage, and in the twentieth century the mature forests have returned, even though the population wishing to use and enjoy them has also expanded.

Today, as from the beginning, the MBL researcher typically first arrives in Woods Hole along with the ferry visitors in the summer. The streets are filled with "ice cream people," dressed in shorts or bathing suits and wandering casually about the streets slurping drippy cones; and "lobster people," whose bright red skin shows that they have spent a little too long in the sun on Martha's Vineyard and who want one last lobster dinner before heading home; and children, impatient to board the ferry because there seems to be little for visitors to do in Woods Hole; and dogs, black dogs.

No helpful signs indicate which way to turn. If one has arrived by bus, the best bet is to walk away from all the people gathered around the ferry dock. If arriving by car, one often waits for the upraised drawbridge (which replaced the old arched bridge around 1910) to go back down and thereby complete the main street over the Eel Pond. The new arrival then moves cautiously down the street toward what *must* be the MBL. Of course, sometimes it is not the MBL because the newcomer first passes the buildings of that other major Woods Hole research center, WHOI (the Woods Hole Oceanographic Institution). Finally, the researcher finds MBL Street and feels secure in the knowledge that this, at last, is the right way. But then, if driving, where does one park? For, since the automobile first arrived in Woods Hole, success has filled the parking lots with cars offering an ever more impressive range of license plates. Somehow, with careful planning, remote parking lots and shuttle buses, and "ethical parking" (don't take even a foot more than you need), they all fit in.

After a student or researcher arrives today, finds a place to stay, and has had time to walk about and explore the landscape a bit, the question asked back home may resurface. Why do biology by the seashore? Why come to a marine laboratory anyway? Why Woods Hole and why the MBL? Some have expressed scepticism about the excuses for going to the sea. The British zoologist E. Ray Lankester, for example, found that "spasmodic descent upon the seashore" particularly suspect and too like a mere summer vacation. Today a few admit that they do not really need to be in Woods Hole to do their work. Some marine organisms can be flown around the country, as witnessed by the number of live lobsters leaving Boston's Logan Airport. But why not work at the seashore? The relative simplicity and diversity of marine organisms makes them particularly useful for understanding life processes. Most researchers really do need to be beside the

Windswept and bare, Penzance as it looked in 1909. Gideon S. Dodds Collection, MBL Archives.

seashore to get the fresh materials in sufficient quantities. The community sense of life at the MBL stimulates the participants. The relaxed atmosphere, the sense of open conversation, and exchange of ideas appeal. Their children love it. Above all, the MBL is a unique center for intense biological research in a community of leading researchers. They all return.

The Origin of the Marine Biological Laboratory

Why was Woods Hole the chosen location for the MBL? Actually, someone who arrived before 1896 would have had to ask "why Woods Holl?"—its official name until that time. Considerable debate about the name—whether named by Norsemen or referring to a hill or someone's family name—circled the small town. Then in 1896 the United States Post Office decided that it would be Woods Hole, much to Whitman's annoyance as he had named several local species "hollensis."

Woods Hole was chosen for the MBL location essentially because of Spencer Fullerton Baird and Alpheus Hyatt, both immortalized by streets named after them in Woods Hole today. Baird served as first (and unpaid) commissioner of the U.S. Fish Commission, as well as secretary of the Smithsonian Institution. He was the sort of fellow who would walk twenty or thirty miles while a young professor at Pennsylvania's Dickinson College just to get a book. When Baird saw Woods Hole in his search for a permanent site for the Fish Commission, he recognized its strengths. As lifelong MBL embryologist Edwin Grant Conklin later put it, Woods Hole had wonderful "natural advantages," namely the "numerous harbors and lagoons, with muddy, sandy, or rocky bottoms, while the coast is so broken by bays, promontories, straits and islands as to afford the most varied habitats." In addition, the tidal currents churn up the food and oxygen supplies in the water and produce bountiful collections of organisms from the nearby Gulf Stream as well as the northern currents. The freshwater ponds provide alternative supplies for materials, Conklin noted. Enthusing further, he suggested that you "add to these things the fact that Woods Hole is readily accessible by rail or boat, that the climate in summer is delightful, the bathing excellent, the mainland and islands charming, the sound with its continual procession of ships always varied and interesting, and you have in Woods Hole not only an ideal place for a laboratory, but also an ideal place for summer residence." Given that Baird had spent some time exploring up and down the coast and that he shared Conklin's enthusiasm for this small town (with its seventy-five buildings in 1871), it should not surprise us that he chose Woods Hole for the Fish Commission operation.

Full of energy and ideas, Baird wanted to build a major research center in Woods Hole. He had local friends who encouraged the enterprise and helped to secure the land, including a plot right across the street from the

◀ *The drawbridge into Eel Pond, with the MBL vessel* Dolphin *coming through.* MBL Archives.

Old Fisheries Building with Crane and Candle buildings behind, 1923.
Norman W. Edmund Collection, MBL Archives.

Fish Commission building intended for a future research laboratory. As it became clear that the Fish Commission could not do everything by itself, Baird encouraged others to join him with a teaching laboratory in Woods Hole.

Baird's friend and Harvard graduate Alpheus Hyatt ran a teacher's school of natural history up the coast in Annisquam. Hired by the Boston Society of Natural History and supported by the Woman's Education Association of Boston, the energetic Hyatt and his assistant and instructor Balfour H. van Vleck had conducted a summer school each year beginning in 1879. At first they met at Hyatt's house, but after two years of that they acquired another site. Sometimes the students proved less than inspiring.

Once Mrs. Hyatt wrote to her husband, who was away at sea with a group of students, that the students left in Annisquam, particularly the women, were dreadful and that both she and van Vleck almost despaired of teaching them anything. But a few prospective professional naturalists also happened into the course, including Thomas Hunt Morgan (in 1886) who later became a major figure at the MBL and the MBL's first Nobel Prize winner.

Hyatt kept the school going, despite discouragements, until the Woman's Education Association decided it was such a success that they should no longer need to fund it. They had always insisted that they would try the school as an experiment, and if it succeeded it should become independent. That time had come. Hyatt was evidently ready for different arrangements. He loved exploring and traveling and was a fine naturalist in the style of the day, but perhaps he had had his share of discouragingly elementary students by then. Besides, the waters of Annisquam were becoming seriously polluted. Baird invited Hyatt down to visit, as he had before, and laid the prospects before him to locate a laboratory and school in Woods Hole. The

Left: Spencer Fullerton Baird, first commissioner of the U.S. Fisheries, 1871–1887, and founder of the Woods Hole Fisheries. Paul Galtsoff Collection of the Fisheries, MBL Archives. *Right: Alpheus Hyatt, first president of the MBL Board of Trustees.* MBL Archives.

Sketch of the laboratory at Annisquam, made in 1884 by an unknown artist. Windpower provided the laboratory with its water for aquaria. MBL Archives.

land next to the Fish Commission remained available. Baird promised that the Fish Commission would help with such potentially troublesome necessities as supplying organisms and running sea water. The Woman's Education Association agreed to donate whatever equipment and materials Hyatt had accumulated for the school. Because Hyatt had already invested considerable energy and money of his own in the Annisquam venture, the new prospects for reinforcements in the form of personnel and funds must have appeared very attractive indeed.

The enterprise took shape, with the expressed purpose of raising "such a sum of money as will secure for teachers of Biology and the general student ample opportunities for the practical study of marine forms. It is hoped that a sufficient amount may be raised to offer additional facilities for original investigation." The group of supporters and future trustees set out to select a site and to raise money for a new biological laboratory on Cape Cod. Woods Hole may have appealed to them for financial as well as environmental reasons since land here was cheap at this time.

In 1887 and 1888, as the planning took place, the town boasted its very own guano factory, an ingenious business that turned local waste products into valuable fertilizers and provided jobs for the townspeople. (The guano came first from local ships returning from the Pacific islands, then from nearby islands.) Begun in 1863, the Pacific Guano Company saved the town from the post–Civil War economic depression suffered elsewhere throughout the United States over the next few years. As one man noted, "the concept, and the product, reeked of ingenuity." That was the problem. Producing fertilizer was a great idea, and using otherwise worthless bird droppings made sense, but it smelled awful. At its height, around 1879, the factory employed 150 to 200 workers and processed 40,000 tons of fertilizer. In 1887, it still served as a strong aromatic deterrent to tourist development, but its end was near. Cheaper fertilizer elsewhere doomed the plant, so that it closed in 1888, causing a local depression. 1887 and 1888 were good times for the wealthy to move into Woods Hole, as the smell was leaving and before the land increased in price and when the economic conditions would have provided cheap labor for household help.

The group from Boston liked Woods Hole. They decided officially to form the Marine Biological Laboratory, with Hyatt as first president of the board of trustees, and to move to Woods Hole. With $10,000 raised over the winter months, they incorporated on March 20, 1888.

They really did not have enough money to hire a director and a teacher. First, the trustees sounded out Williams College biologist Samuel Clarke. He responded to the effect that they must be crazy. "Let us consider some of the points:" he wrote, "the offer is, for me to pay my expenses to Wood's Hole, to continue paying my expenses through the summer and my return expenses home; also to take all the burden of organizing the Laboratory,

Guano factory is to the left in back. The legend on the back of this picture reads: "This point of land, sometimes called Refuge Point, was given to the U.S. Fish Commission by Joseph Story Fay." MBL Archives.

adjusting it to the conditions there and in Boston; the receiving of and answering of applications, the reception and settling of those admitted; the labor of giving lectures, arranging courses and providing material; and the establishment and preservation of a cordial feeling between the U. S. Fish Company and that of the Laboratory—as well as between all those in the Laboratory." He declined.

The trustees then decided to offer the directorship to William Keith Brooks, who was probably the most well-placed biologist in America at the time, with his professorship in zoology at the research-oriented Johns Hopkins University. He had attended Agassiz's Anderson School and had then organized his own Chesapeake Zoological Laboratory for his students at Hopkins. However, one member thought that Brooks might possibly accept the position for no salary and bring the support of Johns Hopkins as well. Optimistically, they offered Brooks the job. Then they waited; and waited throughout the spring and into the summer months. The time came closer to announce and then begin the first summer session in 1888. Finally, Brooks declined. He saw no reason, he explained, to have another laboratory in Woods Hole. The Fish Commission was there. That was good enough for his few professional students. And he really did not see the point of having a teachers' school.

Besides, it was clear from his actions at Hopkins that this shy and retiring man did not want to teach biology to women. And women there would be. That was understood from the beginning. In fact, the original board of trustees included women from Boston. Natural history was regarded as a "safe" subject for women, and women teachers wanted to have experience in the laboratory and the field just as much as men did. As long as the purpose of the MBL was seen as educational and as providing a place for both students and investigators to work together in a community setting, women promised to play a significant role.

Today, women have indeed begun to fulfill that promise. Thirty to fifty percent of students enrolled in courses are women, though a much smaller percentage participate as course instructors or primary laboratory directors. In fact, after the early years the proportion of women students remained well over half until the 1930s, when it dropped off radically, and has gradually risen only in recent decades, as is generally true elsewhere in the professions.

So, having to teach women and teachers, and a strong conservative—or pragmatic—streak kept Brooks from accepting the directorship. The trustees decided to make the next offer to Whitman, the only other American who had directed a biological laboratory, though admittedly an inland facility: the Allis Lake Laboratory near Milwaukee, Wisconsin. Almost immediately, Whitman accepted—at no salary. And he continued as director until 1908, at no salary, and sometimes at considerable expense from his own pocket. Whitman had ambitions to build biology in the United States and to gain a more important professional position than that at the Allis Lake Laboratory, and he had a vision of what biology should be like and what a marine laboratory could do. He worked hard and sacrificed much to put that dream into effect, to create a unique national research place where students, researchers, and scientific leaders could meet and communicate and exchange ideas. He wanted to create a vital organism that would grow and mature. To him the MBL and Woods Hole offered the perfect opportunity.

Modern visitors to Woods Hole often do not appreciate the reasons behind the choices. Indeed most never have any sense of the very real people and ideas behind those street sign names in town: Spencer Baird, Hyatt, Whitman, Conklin, Clapp, Lillie and, yes, even Brooks. Tourists roll through town all summer without ever realizing why all those people are going to the Friday evening lectures and without understanding the striking number of important contributions to biological science that have begun in this spot.

Woods Hole as it looked in 1886, before the MBL began.
Photograph by Franklin Gifford. MBL Archives.

NOTES

One biography of Agassiz is by Edward Lurie, *Louis Agassiz* (Chicago: University of Chicago Press, 1960). Another is by Elizabeth Cary Agassiz, editor, *Louis Agassiz, Life and Correspondence*, 2 vols. (Boston: Houghton Mifflin, 1886); pp. 765–776 discuss the Anderson school and the drain that it caused on Agassiz's life.

A series of pieces in the *Tribune Popular Science* in 1874 details the school's coming into being, as do articles in *Frank Leslie's Illustrated Newspaper, Nature, Science,* and other journals of the day. The centennial program for Agassiz's school, August 13–17, 1973, also included important discussion of the school's history. See also articles by Ralph Dexter, especially "From Penikese to the Marine Biological Laboratory at Woods Hole—The Role of Agassiz's Students," *Essex Institute Historical Collections* (April 1974) 110: 151–161.

The records of the Boston Society of Natural History record the progress of the Annisquam school—its success in attracting students, costs, teachers, problems, and such.

Related discussion of the role of nature study and the place of women in natural history appears there and in most popular science publications of the 1860s and 1870s. A more sustained and scholarly treatment of women scientists is Margaret W. Rossiter's *Women Scientists in America* (Johns Hopkins University Press, 1982); see especially pp. 86–99 on the MBL.

For more on Whitman, see Edward S. Morse's Memoir, *National Academy of Sciences, Biographical Memoirs* (1912): 269–288, and an article by Frank R. Lillie in the *Journal of Morphology* (1911) 22: xv–lxxvi.

Lillie's unpublished autobiography, dated 1926 (?), is in the Lillie materials at the MBL.

Isabel Morgan Mountain has written an unpublished history of her house, on the site of the old Morgan barn. This discusses the fire and Morgan's experiments. She has also generously provided other valuable information and insights during an interview in summer 1987.

Cornelia Clapp on the early years: "Some Recollections of the First Summer at Woods Hole," *Collecting Net* (July 7, 1927): 3, 10.

Beamon Douglas, "My Summer in Wood's Hole," *New York Medical Journal* (1902) 76: 265–269.

Joseph Fay's stories about the possible arrival of Norsemen on the Cape appeared in the *Collecting Net* and in "Track of the Norsemen," 1873 (?). Also recent historical publications about Woods Hole consider such historical stories.

E. Ray Lankester on marine laboratories, "An American Sea-Side Laboratory," *Nature* (March 25, 1880): 497–499.

E. G. Conklin on the advantages of Woods Hole, "The Marine Biological Laboratory," *Science* (1900) 12: 233–244, quotation p. 235.

Baird's interests at Dickinson College are recorded in the excellent archival collections there. On Baird and the Fish Commission, see Paul Galtsoff's *The Story of the Bureau of Commercial Fisheries Biological Laboratory, Woods Hole, Massachusetts* (Washington, DC: Department of the Interior Circular, 1962).

On Woods Hole see Harris W. Clark, "A History of Woods Hole and Adjacent Waters," March 1975, MBL Archives.

Concerning relationships of the MBL to earlier American efforts, see Dexter (reference chapter 1, #6) and articles in the special issue of *Biological Bulletin* (June 1985) 168 Suppl., especially Jane Maienschein, "Agassiz, Hyatt, Whitman, and the Birth of the Marine Biological Laboratory," pp. 26–34.

The letter describing the purpose of the prospective MBL was written by the "Committee on Location" (B. H. van Vleck, W. G. Farlow, and S. F. Clarke), April 15, 1887.

Samuel F. Clarke discusses the directorship in a letter to William Sedgwick, December 18, 1887, and in the MBL trustees' minutes from 1888. Discussion of Brooks's hesitations appears in letters in the Johns Hopkins University and the Museum of Comparative Zoology Archives.

The only summary history of the MBL is Frank R. Lillie's *The Woods Hole Marine Biological Laboratory* (Chicago: University Press, 1944); reprinted in *Biological Bulletin* (1988) 174.

2

Living Here

Day's work over: part of the fishing fleet tied up at Sam Cahoon's dock. MBL Archives.

GIVEN THAT PEOPLE do come to the MBL to stay for a part or all of a summer, what is life in Woods Hole like for them? We could say different things for different people, of course. Some people have had lovely formal homes, complete with servants, while others have been happier living in rustic camp fashion. Some seek privacy away from the lab, others seek continual involvement. Some live outdoors as much as possible, others inside. The wealth of detailed descriptions of Woods Hole life—housing, eating, working, dealing with weather problems, and such—illustrates the richness of summer living in this active small community.

Housing

Now new arrivals no longer have to take their chances scrambling about town for housing when they arrive; there is no hope of such belated success. Instead, they must plan ahead and make their reservations early, often by February if they plan to stay all summer or to bring the family. And they need plenty of money if they intend to rent a cottage in town rather than to secure housing through the MBL itself. Rents for summer cottages have skyrocketed on the Cape in the past decades, especially as new building codes for Woods Hole have prevented much of the sort of wanton and irresponsible growth that has occurred elsewhere along the seashore.

If you were not too fussy, ten or twenty dollars covered a full week's costs nearly up to World War II. In the early 1940s, that time of the three-cent stamp, summer houses sold for about $3,000 to $5,000, while a student could get a week's board at the MBL mess hall for seven dollars and a cot in the simple but adequate Drew House for two dollars per week. No longer do these prices apply! Nor do many people feel comfortable with the sort of musical beds approach that Philip Armstrong, anatomist/embryologist and later MBL director, recalled of his early years. When he was locked out of his room in town, he headed for the men's dormitory and plopped down on any empty bed to sleep. When the occupant showed up, he simply moved on to another. Others slept outside, on the roofs. People have since then acquired a taste for privacy and also for baths and showers, which were all rare in those early days at the MBL when even running water remained a treat.

Since those early years, the MBL has made an effort to provide housing for its community, with the trustees perpetually concerned about how to raise funds to provide sufficient dormitory space for students. At first students and investigators alike boarded out in town. Then the development of what had once been the small island of Penzance Point, after the demise of the guano works, provided opportunities for those who could afford the property in the growing resort community. Various other arrangements have arisen to provide housing as well. One woman recalled when the men students roomed on the lower floors of one house, and the women on the top where they could dry their hair and hang up laundry in the sunny cupola atop the building.

For a while in the early 1900s students found themselves rooming in the old stone Candle House. The "Hotel Majestic" they called the structure, though it smelled a bit odd and clothes had to be hung from rafters for lack of closets. The building is beautiful now that it has been renovated and turned into a modern and even air-conditioned administration building. In the bunkhouse days that beauty was harder to see. Built in 1836 at the height

"Hotel Majestic," summer quarters for many students in the early days.
MBL Archives.

At the fish trap, 1909–1910. Torpedo-shaped car in the foreground was used for bringing back the squid. It was towed behind the collecting boat and served to keep the squid alive. Gideon S. Dodds Collection, MBL Archives.

of Woods Hole's whaling activity, the building had housed the quantities of whale oil brought in to make spermaceti candles, before the discovery of oil in western Pennsylvania and before gas lamps became popular in the 1850s to replace whale oil as a source of lighting. The smell of whale oil lingered into the bunkhouse days, and at times was tinged with the odor of the formalin from the supply department downstairs in the basement.

Whaling and Fishing

Whaling and fishing had been prominent parts of Woods Hole life before science and tourism came to dominate. Though the town never offered as major an operation as Nantucket, Woods Hole did send out its whaling vessels to collect oil in the early and mid-nineteenth century. After the MBL began, fishing played an important part of Woods Hole life, as it did along much of the New England coast. Old-timers and not-so-old-timers recall fisherman friends, often extremely well-educated and fascinating people, coming over for an evening chat after a good day and bringing along a collection of quahogs, scallops, fish, and even a few odds and ends that they thought the biologists might like—either to eat or to study.

New Yorker George Scott recalled one time when friends took him along on a trip to do some bottom fishing. As Scott reported, "That trip gave us everything: Hove to for a day in rough weather, and fishing ruined for several hours by aggressive sharks tearing a hole in the bag of the net in the trawl as it was brought in, letting out the fish which were devoured voraciously."

Whaling provided good stories also. In 1899, Ulric Dahlgren (a Princeton University histologist apparently of Viking descent) organized a swordfishing party. The vessel *Vigilant* had a traditional "pulpit" at the bow for watching the fish, and the remaining necessary equipment. Dahlgren made up a party including embryologist Edwin G. Conklin, cytologists Edmund B. Wilson and Thomas H. Montgomery, Japanese artist K. Hyashi, and a few others. Expecting a two-day trip, they loaded food and water and set off for the open ocean in the light wind. As the wind picked up, they sailed around Gay Head and out to sea, accompanied by a rosy sunset. After a night without much sleep, the group arose for a breakfast of bacon, eggs, hardtack, and coffee, then sailed on. Soon came the excitement they had hoped for. Spotting some sort of cetacean in the distance, they moved toward it. When they got close enough, Dahlgren speared the creature with a dart and iron but not the harpoon. It turned out to be a whale. The whale was not happy and thus dived into deeper water, returning in a few moments.

Dahlgren decided then that he would try to wear out the whale, and he put out a dory. Two men set off in it, with two sets of oars and a 300-foot piece of rope. They caught up with the barrel attached to the dart and iron and grabbed hold. Securing the rope to their harpoon rope, they were thus tied to the whale. The whale headed at full speed for open ocean, with the dory and two biologists attached. Normally calm and collected, Wilson stood on board shouting "I'd give a hundred dollars to be in that boat." But then the *Vigilant* fell behind and lost sight of the disappearing dory. While helplessly waiting for the men and dory to return, those remaining set about to catch the swordfish they had intended to capture, but instead harpooned a giant marine sunfish, *Orthagoriscus mola*. With somc difficulty they wrestled that fish on board, which distracted them for a while. They then finally recalled their missing friends, who had disappeared more than two hours before.

The men resolved to set out in the second dory in search of the first, because the *Vigilant* was completely becalmed by then. But they had no oars left. The first dory had left with both sets. They tried boards, which did not work. Concerned about the two men with no food or water out in the hot sun, they tried frying pans as oars, but the handles were too short. They then found two brooms and set out, "sweeping the Atlantic." Fortunately, the other men returned to sight and the sweepers could give up their absurd effort. The whale had moved around so much with his diving and resurfacing that the rope had frayed and broken. Dahlgren estimated that they had been towed about ten or fifteen miles before that though. As Conklin reported, their evidence suggested that the whale was a finback, and probably even Dahlgren would have failed to wear him out.

This "fishing expedition of a lifetime" all ended by 11:00 A.M., much to everyone's astonishment, as they observed that people back in Woods Hole would just be going to church. As a student at Hopkins, then a professor at University of Pennsylvania and later head of biology at Princeton, Conklin loved telling stories: especially this one, where he stalked up and down the beach wielding a harpoon. He also observed that the group never quite wanted to embark on such an adventure again, even though the sunfish provided a great wealth of parasites for a number of research projects back in the lab. They all felt that a second excursion would be anticlimatic.

Commercial Resources

The more normal fishing enterprises also provided fresh fish in the shops that used to line Water Street. Now, as in most of America, MBL folk generally hop in their cars to drive to the nearest town, Falmouth, for the bulk of their groceries, buying only smaller items in the small Woods Hole store. But within the recent memory of many, Woods Hole housed a meat market and several groceries, which provided a variety of fresh food. Sam Cahoon's fish business provided dockside seafood for the village, "a place of distinctive and pleasant smell, which was a blend of fish and smoke and

Behind Sam Cahoon's fish market on the dock, where much of the processing took place, circa 1930. Photographs by Alfred F. Huettner, MBL Archives.

steam," the latter two from the train yard next door. Once a popular actress was in Woods Hole for a visit. She rowed over to the town from Penzance Point, the fashionable lower tip of Cape Cod, to get some fish; but she had forgotten her money. She announced that she was the actress Katharine Cornell, expecting her familiar name to be sufficient. The generous and honest Sam Cahoon responded that, in fact, he had never heard of her but that did not matter to him. "Give the lady her fish," he said.

In addition to fish markets, the town had its drug store, dry goods shop, and even its own hardware store, the latter run by Edward Swift. The popular Mr. Swift recalled when the whaling ships were built in Woods Hole, as well as when the first train had arrived and when the last train left. In addition, a few informal businesses kept things happening at times. Rumrunners during Prohibition added excitement, especially when they got news of a coming raid and had to toss the wooden crates overboard for later retrieval. Town boys tried to get there first. Woods Hole was a busy town, with an active commercial life and a scientific population growing side by side.

World War I brought changes in Woods Hole. The Great War ended with a heroic burning of the Kaiser at a grand bonfire, very inspiring to the local youth. During the war, Woods Hole had a few exotic exiles, including one Russian family who built their own house by hand, and the future president of Czechoslovakia. Following the war, people began to build more houses, sometimes ordered prefab from Sears and Roebuck. Even this did not provide for the influx, however, so that during the 1930s the younger researchers found it difficult to find a place in Woods Hole and increasingly looked elsewhere, as others have done more recently for the same reasons.

The second world war brought its own post-war building expansion and relative financial security so that more people could afford to buy summer housing, or even permanent housing. Nobel laureate Albert Szent-Györgyi, for example, arrived during this time (in 1947). He had visited the lab in 1929 but would not have recalled it in particular, he said, because he had visited so many marine laboratories and had not spent long in Woods Hole. Except that his hosts then had prepared a clambake, complete with lobsters, for the distinguished European visitors who made their way down the coast from the International Congress of Zoology in Boston. When

Albert Szent-Györgyi on the beach at Penzance. MBL Archives.

Szent-Györgyi suddenly had to leave Hungary in 1947, ten years after having received the Nobel Prize for his work on vitamin C, he recalled the lobsters and the lab. Here was a laboratory where he could rent research space and not have to owe anybody anything. He moved to Woods Hole for good and made a great impact on the MBL. As recent director Paul Gross put it, "During the height of his career, when I was a student here, he was for many of us the paradigm of the scientist as the spearhead of culture." He was, in short, "the very model of a man."

Szent-Györgyi rented a house for a year, then decided to buy. The real estate agent was not much more optimistic than a realtor might be today, insisting that there really was nothing to look at. Szent-Györgyi persisted, so the agent drove the car around to point out the remote possibilities. When he saw a boarded-up house on Penzance Point that he simply had to have, the agent insisted that it was too big. In addition, Szent-Györgyi's money was all tied up in Hungary. He bought the house anyway, complete with furniture, lovely china for twelve, glassware, and silverware, all from the very gracious and generous former owners. Many MBL houses have been acquired in similarly unorthodox manners, though seldom so fully equipped and with so little immediate money down.

Yet that increased desire among scientists to own houses also put pressures on the existing housing, so that there were far fewer places to rent in town. The MBL was going to have to provide for its visitors. Today those visitors may find themselves in any of an increasing number of rooming houses owned by the MBL, and embryonic plans for a second modern dormitory building are taking shape. Some of today's housing is in noisy old wooden buildings with rooms rambling everywhere, so typical of remodeled Cape Cod houses. Others are brick buildings with rooms arranged along the long hallway in 1920s fashion, when the brick apartment building was constructed in 1926 to meet the perennial housing shortage. Inevitably, these will probably make way for larger and more modern facilities that will use the space more efficiently or provide more laboratory space as well.

Some visitors, including those who arrive for shorter times to use the library or to work briefly in someone else's laboratory, find themselves assigned to the more modern Swope Center, built after the post–World War II "marriage boom." Swope features private bathrooms and a hotellike atmosphere, quite a luxury compared to the old wooden dormitory days.

Swope also has a built-in wake-up calling system, despite its absence of telephones. The bell tower next to the Eel Pond and associated with St. Joseph's Church rings the traditional calls at 7 A.M., noon, and 6 P.M., thanks to Frank Lillie's wife, Frances Crane Lillie. In 1929 she presented the tower, with its small chapel and garden areas, to Woods Hole, calling for the regular ringing of its two bells, Mendel and Pasteur, The Mendel bell—

The bell tower. MBL Archives. *Richard Huettner, son of the scientist-photographer, standing in the doorway of the bell tower, 1930.* Photograph by Alfred F. Huettner, MBL Archives.

representing the patient biologist who carefully counted his sweet peas in the monastery gardens and who saw the basic foundations of genetics forty years too soon—rings first, calling "I will teach you of life,—and of life eternal." Pasteur, French chemist, microbiologist, and vanquisher of materialist ideas about spontaneous generation of life, responds, "Thanks be to God." The 7 A.M. call gets everyone to the fixed-schedule breakfast on time and back to work for another busy day.

In the increasingly expensive and tourist-influenced Woods Hole Village, providing housing is essential for the MBL to attract researchers and students. To attract instructors for the courses and advanced researchers with children in a time of relative prosperity, where people are not used to the crowded quarters and primitive conditions that most visitors endured in the early years, the lab found it had to provide real housing instead of just rooms. Thus, over the years, houses have been added on the old Fay property or in Devil's Lane.

In 1916 the Fay family, always friends of the laboratory, sold twenty-one acres of wooded, hilly land to the MBL at a very favorable price. Called the "Gansett property," this land was laid out in lots with the understanding that buyers must give the MBL first option on resale and that the trees must be preserved. The Devil's Lane property, acquired in 1925, had similar conditions, though it originally included land that some thought might best be used in the future to develop chemical and physical laboratories to operate alongside the biological lab. The MBL sold parts of the Devil's Lane and neighboring properties acquired later and has built some housing there itself for rental purposes. As recently as 1986 twenty new houses joined the earlier cottages at Devil's Lane. Distinctly designed to remain inconspicuous

Some of the new houses built for summer investigators, 1986. MBL Archives.

The Children's School of Science on MBL dock, 1953. Photograph by Alicia Hills. MBL Archives.

and essentially anonymous, they remain simple but adequate and have taken considerable pressure off the already crowded facilities.

In both the Gansett and Devil's Lane areas, as in much of the rest of Woods Hole, the designers named streets after the early scientists. In 1986, MBL biochemist Seymour Cohen and Ruth Gainer piloted a project, carried out by a group at the Children's School of Science, of providing a historical guide to the people behind the street names in Woods Hole.

The Science School

The Children's School of Science, known informally as the Science School, is a unique Woods Hole institution. Housed in what was until very recently the Woods Hole public school building, this summer school attracts children back year after year. Special courses for different ages, a range of field trips, and an informal atmosphere make kids want to go to school during summer vacation.

The Science School began in 1913 with the efforts of Mrs. Wilson, Mrs. Morgan, Mrs. Lillie, Mrs. Warbasse, Mrs. Conklin, Mrs. Calkins, Mrs. Crampton, and others—a virtual who's who of MBL wives. They wanted to provide summer supervision and activity, including singing and dancing, but also an introduction to science, with nature study and even research. The women joined in the organization and developed planning committees until they hired a proper teacher for the program which opened for business in 1914. Every year thereafter the school has had a full schedule, with the exception of 1916, when a polio epidemic kept it closed. The group decided not to open at all that year, though they paid the teacher anyway. A few people felt that they should close during World War II, but the leaders decided that however few children there were, then more than ever they needed some stability and activity. An impressive number of Children's School alumni have gone on to such careers as scientific research, teaching, or medicine, happily encouraged by the school.

The summer school accumulated quite a nice library as well, for which the regular winter administrators hesitated to take responsibility. So each year, most of the summer school books were taken off to the Woods Hole community library, once called the Social Library. Just a few remained behind for the public school teachers to use during the year. The school had eight grades for a while and then six. As Woods Hole children grew up, they took the train (and later the bus) to the Falmouth school each day. Now all the students commute to Falmouth for their full school careers.

Food

Returning to fundamental matters: the families, especially summer visitors, must consider such obvious questions as where to eat. Those with cottages can cook at home. If, that is, they have a car, are willing to walk or take a bus to Falmouth, or like the limited diet allowed by the indispensable Food Buoy on Woods Hole's main street. They can enjoy the restaurant offerings in town, or purchase a meal ticket or a single meal at Swope.

Not so many years ago, the majority of the community still ate at least lunch in the dining hall. Even many people with houses often ate "in." The community was together then, and the students mixed comfortably with instructors and the occasional Nobel laureate. Not so much any more. Students eat there, along with those researchers who are staying only a short time or who are single or who are staying in the dorms without cooking facilities—or those rare, delightful few who want to carry on the community experience of older times when everyone met over meals. The

The Mess, circa 1914. Entitled "Dedication Day" by the photographer. It was on July 10 of that year that the first permanent building, Crane, was dedicated. Ida H. Hyde Collection, MBL Archives.

The Mess in 1953. The face had changed, but the Mess was still used much as before. Photograph by Alicia Hills, MBL Archives.

Inside the Mess, 1953; Mr. Martin checking tables. Photograph by Alicia Hills, MBL Archives.

existence of so many alternatives elsewhere draws busy people away from what used to be a life center of the MBL.

During the time of Old Main, the original set of attached, simple gray wooden buildings built one at a time in 1888, 1890, and 1892, most people at the MBL ate at the Mess. One young wife recalled that other women were astonished that she chose to cook at home rather than to join the others at the convenient Mess. She responded that her husband had married, in part, precisely to get such home-cooked meals. He was not fond of the institutional food. Others recalled the food as very good in those early days, especially the inevitable Saturday night baked bean dinner. Originally housed along with several rooms for investigators in what had been the Fays' gardener's cottage at Little Harbor, the Mess soon moved nearer to the main building. Rebuilding the Mess became a priority when it was destroyed by fire in 1920.

The growth of the population demanded the flexible approach of the Mess, with everyone eating together, family style. Sometimes notations on napkins attested to the exchange of scientific ideas there. After meals, people would continue the conversations begun inside. This was a time for people to break from the individual research and to enter the community, to eat in leisure and then to gather on the porch talking, to bring the parts together into the living MBL whole.

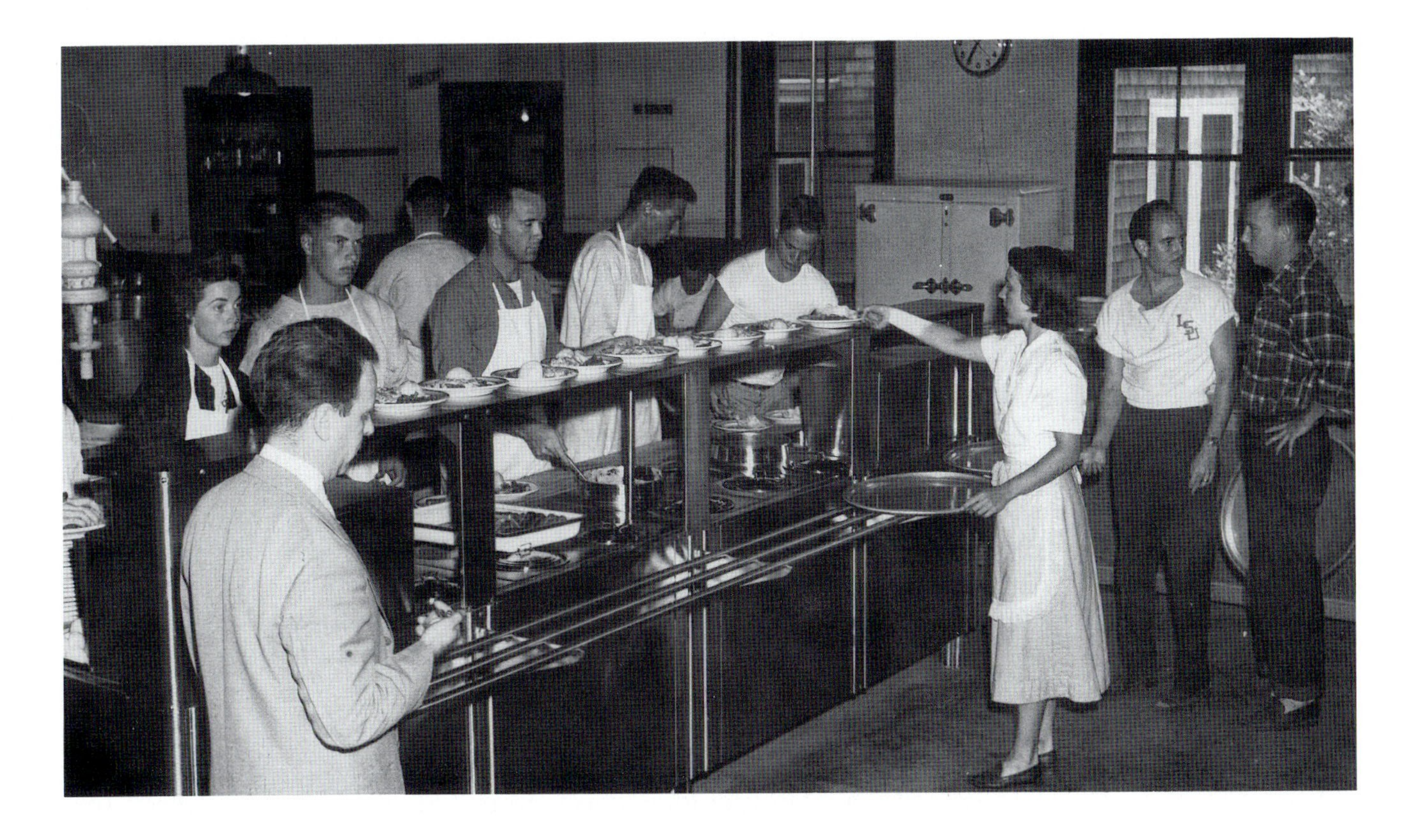

Modernized service at the Mess in 1953. After Miss Belle's retirement, meals were served, no longer by waiters, but cafeteria-style. Photograph by Alicia Hills, MBL Archives.

Once the Mess building was built, it served as the central meeting place until replaced by Swope. Only once did the MBL crowd gather somewhere else to eat for any length of time. During World War I, the Navy took over some of the MBL's buildings, including the Mess. Woods Hole was a good site because it already had the government's Fish Commission buildings and the deep water that Baird had long ago selected to enable the Commission's large ships to dock. In addition, the small town at the bottom of the Cape was relatively isolated and secure. Yet when the Navy moved in for the winter of 1917–1918, the MBL trustees vowed that they would make all efforts to keep the place running the next summer. They believed that "it is the duty of scientific men not called directly to war service to maintain their scientific activities."

In 1943, the Navy returned once again to set up a temporary base in Woods Hole. They took over the old lecture hall, the old botany building, the apartment house, and Penzance Garage on the wharf, in addition to the Mess. MBL arranged for alternative boarding at the Nobska Inn, now the Woods Hole Inn, which was not really large enough to hold everyone at once. In addition, the general rationing and shortage of meat and butter made the fare less satisfactory. Because somewhat less than half the usual number of researchers and students had come to the MBL for that wartime season, the effect was less serious that it might have been. Yet Woods Hole old-timer Dot Rogers recalls that she had the only car around and had to make many emergency trips to Falmouth hamburger stands to keep people happy. Fortunately, the lab managed to sustain the usual assortment of collecting trips, Friday evening lectures, and library facilities, even while researchers ate elsewhere and had to forego special seminars from visitors who could not travel without rationed gasoline.

At an earlier time, the Mess had been rather more formal. Each table had a host, and the same group sat together through the season. The tablecloths remained in place through the week. One host devised a plan whereby anyone who spilled food onto the cloth would have to cover the spot with an appropriate coin. At the end of the summer, they took the collected money and had a fine party for all. Later Harvard biologist George Howard Parker and his companions raised the ante so that the spill had to be covered by a dollar bill, with the money used to buy jam and other luxuries—even one season hiring an organ grinder, according to some perhaps fanciful reports—to supplement the substantial but uninspired mess.

"Miss Belle" (Downing) ran the Mess for many years after 1940, and insisted on white linen tablecloths and fresh flowers every day, even if she had to go out and pick daisies herself. She determined where people would sit. She mixed the senior scientists and the younger researchers and students all together at a table, including developmental biologist Ernest Everett Just, who was one of the very few black biologists at the MBL. This caused quite a shock for one southern student, who recoiled at the expectation that he would have to wait on a black man at dinner; it just felt wrong. Then when Just asked him what he wanted to study and invited him to visit his lab, his hesitation passed and his prejudice dissolved. No one thought of moving after Miss Belle made her assignments, so people sat with the same group long enough to get to know them. And they met people from all different aspects of MBL life.

Students earned enough money to pay their expenses in courses by waiting on the tables in those days before the disruption of war; eventually economic necessity turned the operation into a cafeteria. Singing has always played an important entertainment role at the MBL, and waiters found themselves immortalized in numerous offerings. Even Alfred Romer, later director of the Museum of Comparative Zoology at Harvard, and Yale biologist J. P. Trinkaus, an MBL researcher for nearly fifty years, earned their way at the MBL as waiters. One set of verses to the tune of "John Brown's Body" went:

Johnny was a waiter at the Mess of MBL
And the trials of occupation there would be hard to tell
It was a busy life with him when everything went well,
Alas when it did not.

CHORUS: *Oh, the joys (trials) of being a waiter*
With everybody late or later,
While their appetites grow greater
At the Mess of MBL.

Early every morning when his dreams were at their best,
The alarm would sound beside him and disturb his peaceful rest
And with an awful effort he would manage to get dressed
and eat his breakfast first.

Then he would take his stand beside the table with his tray
And carry in the mackerel for breakfast every day,
And in and out the dining room would swiftly make his way
To bring the people food.

Trinkaus points out that the waiter had considerable power since he could give the obnoxious people cold coffee or burned toast.

Today, the dining hall in Swope has no morning mackerel, as it once routinely did, nor tablecloths nor waiters for everyday meals, and no assigned tables. No Miss Belle reigns in Swope. Whatever the advantages of the more freewheeling system today, it is clear that something has been lost socially. Today people may spend ten minutes, eating silently, or the students may eat together at one table and the instructors together at another. Social mixing at the MBL has become more haphazard as the style has changed and the community has grown. Even in the winter, when storms batter the Cape and occasionally close off the bridges and shut off the population—even then people have lost the habit of comfortably sitting down and talking with people they do not know.

Weather

In September of 1938 the Mess was closed for the winter season as usual, but the MBL had not yet completely shut down. A few people remained working in the laboratory. On the 17th they heard reports that a tropical storm was headed from the Bahamas toward Florida. That storm then moved northward on the 20th, toward Cape Hatteras, bringing high winds and water. The conditions would deflect the storm out to sea from North Carolina, the weather bureau reported. By midmorning of the 21st, the weather station in New York realized from observing the rapidly dropping barometer that a major storm was on the way north rather than out to sea. But they had no authority to issue official warnings. Only the Washington office could do that, and their information indicated no cause for concern.

With no preliminary warning of more than a seasonal gale, Woods Hole did not worry. By 4 P.M. when the electrical power went out in the labora-

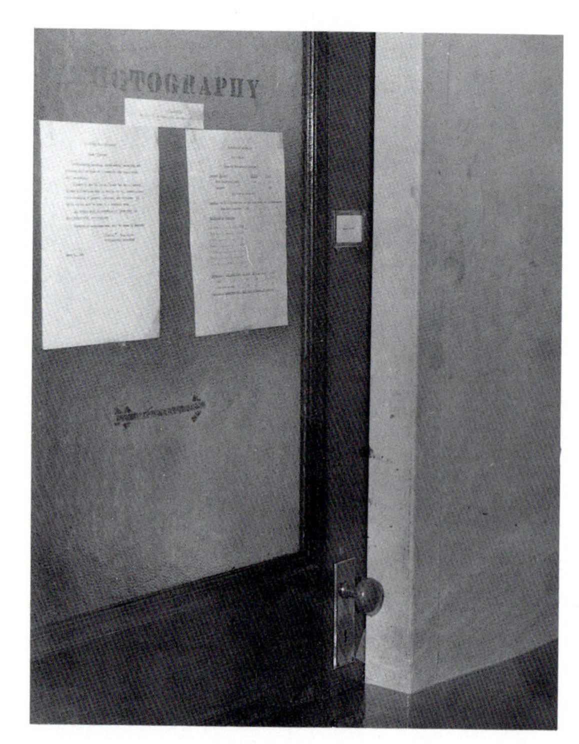

The basement of Lillie, flooded during Hurricane Carol, 1954. MBL Archives.

tories, notifying the scientists and other workers of something unusual, the water had already begun to rise rapidly. Over the next few hours a storm wave brought water rushing over the breakwater and into the Eel Pond next to the MBL's main buildings. As the pond rose to levels no one remembered having seen before, streets flooded and the supply building filled with four feet of water.

Maria von Bertalanffy recalled even thirty years later the horrors of that day. She and her husband, Ludwig, were working in the lab when they noticed the storm suddenly brewing. They went outside to take pictures of the dramatic skies and rising waves. Then they noticed that water was actually rising in the streets and that their rented room on the first floor of a house on Main Street might very well get wet. They decided to go home to move things to safety. But as the winds and water rose rapidly, Maria took the camera back to the lab first, intending to join her husband at home as soon as she could. She returned to the flooded spot on Main Street and began to cross the rising waters. As she struggled a voice urged her back, but she ignored it. Then a powerful hand grabbed her, and the Coast Guard man ordered her back to the laboratory building. A dozen or more people huddled together there, each with his or her own worries. Soon someone came and told them they would have to leave, to go across the street. Evidently batteries in the basement had gotten wet and were emitting what were thought to be toxic fumes. The group clung to each other to struggle across the street against 100-mile-per-hour winds and waited tensely in the familiar Mess building well into the night for relief from the storm and word of relatives and friends. News of drownings and of lost houses and boats began to come in. Finally, relief arrived when the wind picked up from the north; water flowed from the Buzzards Bay side through the streets of Woods Hole and out to sea. Maria's husband arrived safely, at last. Then it was time for everyone to assess damages and to compare stories.

Water had rushed into the basement of the brick buildings. Chemicals of all sorts floated around chaotically in the dark, deepening water. The carefully gathered animals in the collecting cage next to the dock had escaped as the high water released them. The reprints in the library basement absorbed water, turning an estimated 15,000 to 25,000 of them into a worthless, soggy mess. Main Street had had knee-deep water, and the drawbridge foundations had shifted and cracked with the flood. But the rushing water from Buzzards Bay pushed back to safety many of the boats that had nearly been beached on Woods Hole streets.

In total, the MBL suffered only about $20,000 damage, and the Carnegie Institution, which had befriended the MBL before, quickly came forth with the money to make repairs and replace supplies. But outside the MBL there were drownings and great financial loss. Individuals lost their boats or suffered damage to the contents of their basements. There was also such great devastation along the train track into Boston that no one around at that

Receding water in Eel Pond after Hurricane Carol had passed, 1954. MBL Archives.

time would ever forget it. The MBL bricked up windows in the basement to keep out the flow of water in future storms, and the walls are waterproofed periodically. Yet water can still rush right through the bricks in high winds. Pictures hanging on many Woods Hole walls remind people that things have not always been so tame as they seem on a peaceful summer day.

Looking toward Penzance Point during 1938 hurricane. F. R. Lillie papers, MBL Archives.

NOTES

Material from Philip Armstrong, George Scott, and Albert Szent-Györgyi comes from taped interviews at the Woods Hole Historical Collection (hereafter: Historical Collection). George Scott, Historical Collection interview, recalls the fishing adventure.

Conklin was a marvelous storyteller who retold his stories many times, so that a number of people cite them. One published version of Dahlgren's whaling expedition appeared in "M.B.L. Stories," *American Scientist* (1968) 56: 121–129; "The Whaling Expedition of 1899," pp. 121–126.

A number of people told me that Cahoon's was next to the train depot. Similarly, many people in personal communications recall the story about the popular Sam Cahoon and Katharine Cornell.

Woods Hole during the wars is especially remembered in the interview with Robert Kahler, Historical Collection. Newspaper clippings and other notes of various sorts in the Archives confirm the impressions of what life was like during both wars, as do MBL trustees' minutes and annual reports.

Paul Gross, "Report of the Director," 1986 annual report, *Biological Bulletin* (1987) 173: 43.

Discussion of the bell tower and of other details about the building and development of the MBL are included in the MBL trustees' minutes and the annual reports. See also *Collecting Net* (July 6, 1929).

Most unpublished and published accounts of the MBL prior to World War II, as well as the various interviews, mention the Mess and its importance as a central place to gather, to see everyone, and to talk interminably about science. Dot Rogers and Donald Lahy recalled events related to the Mess during an interview to clarify details in summer 1987.

Songs come from old MBL songbooks and single items, largely undated, that people have sent to MBL over the years.

An interview with Elsie Scott, Historical Collection, provided information about the Children's School of Science, as did numerous items in the *Collecting Net.*

Trustees' reports for the war years record the official actions.

The *Falmouth Enterprise* discusses hurricanes. Also, on the hurricane of 1938, Maria von Bertalanffy, "Hurricane in Woods Hole: Thirty Years Ago," in the MBL Archives.

3
Buildings and Budgets

Aerial view of Woods Hole showing the buildings of the three scientific organizations: Fisheries, MBL, and WHOI. MBL Archives.

The Swope Center, which opened in 1971. It houses up to 168 people in its 84 rooms, and the dining room seats 362 at one time. Photograph by F. P. Bowles, MBL Archives.

THE PERSON ARRIVING at the MBL first encounters Swope, the check-in center for the entire laboratory as well as a major dormitory and dining hall. Until 1981 check-in took place in the Lillie Building, and if one arrived after 5 P.M. or on a weekend, he or she had to wait patiently until the watchman returned from his appointed rounds. It is still easiest to arrive during a weekday.

Buildings Old and New

The short-term or winter visitor will usually stay in the modern Swope Center. With the increased housing demand after World War II, the MBL decided to build a new dormitory. After various disagreements about location and several changes of plan, MBL friend Gerard Swope, Jr. purchased a piece of land on the Eel Pond. The replacement of the "Do-Re-Mi" houses there with the large, concrete Swope building in 1971 made a major change in the landscape of Woods Hole. As one nostalgic old-timer griped, the place was getting to be less unique and more like any old ordinary laboratory.

Students enrolled in courses congregate mainly in the Loeb and Whitman buildings next to the original site of Old Main. Landscaped with outdoor tennis and paddle tennis courts, these buildings provide modern lecture and laboratory space. The teaching laboratories are large, well-equipped rooms not unlike the original general laboratories except in detail and equipment.

There are advantages to the relatively modern design of the newer buildings. For example, in the 1940s and 1950s the older buildings caused trouble for the unsuspecting researchers who had begun to embrace the new and exciting work with radioactive isotopes. The isotope containment room where researchers kept experimental specimens was on the south side, which climbed in temperature as the summer afternoon progressed. The researchers kept their control samples back in their labs, away from

Do-Re-Mi houses on the left, circa 1929. Photograph by Matthew Steiner, MBL Archives.

the isotope room so as not to contaminate them, but many of their laboratories were on the much cooler north side of the building. People obtained shockingly unexpected results until someone finally noted that it was the five or more degrees Fahrenheit of difference in control and experimental samples, due to the sun, that was producing the unexpected variations. It took quite a bit of wasted research during that time to teach some of the enthusiastic isotope physiologists to take better care in controlling environmental factors. The new buildings have solved many of the most obvious problems while maintaining the characteristic unpretentious MBL style.

They have also managed to avoid the fire hazards that characterized the early wooden buildings. MBL worker Robert Kahler recalled a time when he was a boy and his family was returning from a picnic. They saw men on a roof, pulling on a hose reel and trying to put out a dramatic fire in the barn belonging to Columbia University embryologist Thomas Hunt Morgan, which was located on a lot down Buzzards Bay Street from the

Equipment used in 1955 to measure the electricity produced by the squid neuron by means of radioactive tracers supplied by the Atomic Energy Commission. Photograph by Fritz Goro, Life Magazine.

Morgan home. The neighbors ensured that the rented cow escaped, he recalled, though someone else pointed out that the experimental mice did not fare so well. A cow seemed more important to the town residents.

Another time, the neighborhood awoke to the smell of smoke and the flurry of activity as people rushed about. The sky glowed red from the flames when the Mess burned. Everyone went to work to keep the burning and flying embers from catching the whole town on fire. Even up on the golf course, long glowing embers found their way from the intense fire.

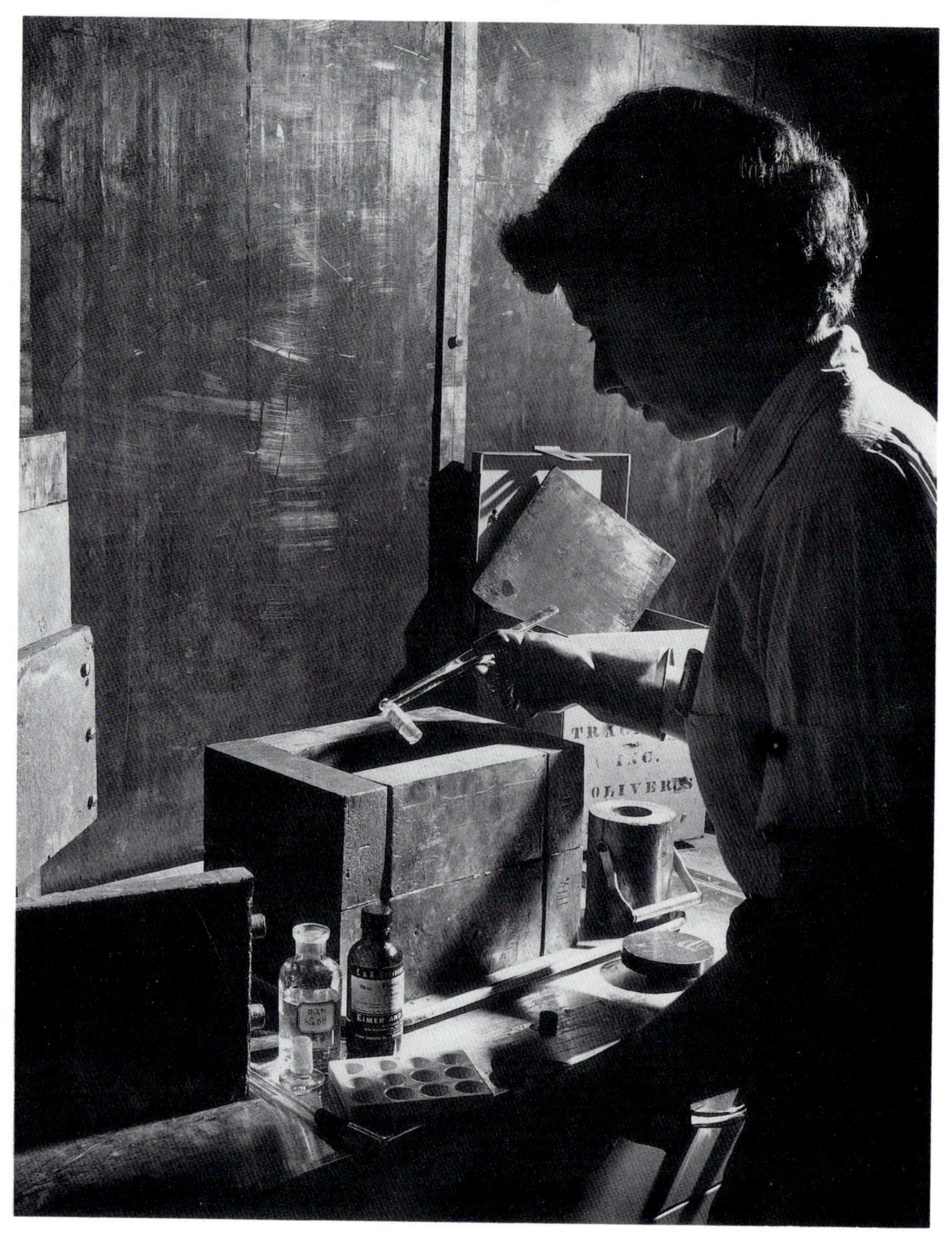

The use of radioactive tracers. In 1955 this was so newsworthy that even Life magazine took an interest. Photograph by Fritz Goro, Life Magazine.

Evidently, oily rags in the Fish Commission's machine shop had started burning; then there was nothing anyone could do until the brand new fire truck arrived. The water pressure was low, and even when the truck came, they kept it at a distance because it was so new and nobody wanted it to burn. Fires have become considerably less of a peril in today's brick and concrete buildings. Attention has been paid to protecting laboratory space in particular.

Instructors in the courses generally receive their own laboratory space as part of their compensation for teaching. As invertebrate embryologist Winterton C. Curtis noted when he was asked to travel from the University of Missouri (not so easy in those days) to teach the invertebrate course in 1908, "the call of the MBL was always 'the call of the blood' for me. Then too," he added, "I needed the money." Not that very much money has ever been involved. Access to lab space clearly offers an extremely important incentive. It not only eliminates a major cost, but it also provides the opportunity for instructors to pursue their own research while teaching a dedicated group of highly selected students.

At first, the MBL followed Baird's model and the Naples model of encouraging subscribers to the summer labs. A university would attain the status of cooperating institution by paying for a table or a laboratory, then send one or more researchers to occupy it. This meant the same labs for the same people year after year, providing reassuring stability. With time, more institutions such as Oberlin College set up scholarships allowing them to send a student or investigator each year. Some Catholic orders reserved space to send a priest or a nun each year to work at the MBL. The MBL allocated space to those institutions that paid without regard for who could use the space to best effect. Many of these programs have been discontinued in recent decades for various reasons, including the MBL's desire to exert greater control over who uses its limited resources. As a result, since the 1950s, lab space has often been assigned on an economic basis. If you can pay, you have a lab. If you can pay more, you have a bigger or better-located lab, perhaps in the Whitman or Loeb Building. How to deal with the requests, which often surpass available lab resources, has become a problem for the administration and the Research Space Committee.

Though many of the new students have no idea who Whitman and Loeb were, it is fitting that these more modern structures carry the names of men who cared about the educational function of the MBL as well as about research. Whitman always demanded that the MBL must include a mix of investigation and instruction, whether in elementary natural history or more advanced techniques and methods. The auditorium in the Whitman Building holds lectures for classes and special weekly seminars on neurobiology, cell biology, and other topics of wide interest, as well as, more recently, public lectures in the history of science and discussions of science writing. Whitman would have approved.

Physiologist Jacques Loeb also would have liked the fact that his building combines research and course work. Although personally controversial in many ways, as historian Philip Pauly has recently shown, Loeb also had a well-documented dedication to introducing sympathetic and committed young investigators to research. As the first instructor in physiology at the MBL—indeed Whitman enticed him to set up the entire physiology

Jacques Loeb with his family at their home in Woods Hole. Conklin Collection, MBL Archives.

department at the MBL—Loeb had the opportunity to introduce students to new and exciting research not normally part of zoological or natural history programs. One Woods Hole native recalled the times that Loeb impressed the youngsters by heating glass over a bunsen burner and then bending it.

By the mid-1890s, Loeb's study of the effects of altered seawater salt concentrations and of artificial parthenogenesis attracted wide attention; by 1900, even receiving consideration in the popular press as well as professional journals. Students wanted to learn how to "control life" as Loeb had done by artificially stimulating the beginning of the fertilization process with chemical manipulations. One enthusiastic beginning student went to Loeb "with more zeal than knowledge" and reported that in the sea urchin *Arbacia* he had discovered that the eggs had an interesting ability to repel other organisms. Loeb explained to him that the eggs have a layer of nearly invisible jelly around them. He suggested that the young man apply india ink to dye the jelly and make it visible. The next day, the enthusiast returned with a marvelous discovery: india ink produces artificial parthenogenesis (or makes the egg act as if it had been fertilized by sperm when it had not). Loeb patiently pointed out that the fellow must have diluted the powdery ink in tap water—which *does* produce parthenogenesis because it changes the salt concentration of the water to an abnormal level. To Loeb india ink had meant the powder; to the student, the liquid. That young man reportedly learned to reflect more deeply about the hidden factors that might not seem important but might turn out to be the keys to a major discovery. Loeb helped to turn many eager students into scientists.

As his student (later University of California professor) W. J. V. Osterhout recalled, many students shied away from the sharp criticisms Loeb offered at times but when they did finally approach him "they were won over by his kindness and anxiety to be helpful." The teaching laboratory building at the MBL is therefore appropriately named after this outstanding scientist, even though he spent much of his life in a research, rather than a teaching, position.

In fact, the current Loeb Building lies across the street from, and replaces, Loeb's own laboratory building, built and equipped for him by the Rockefeller Foundation after he went to work at the Rockefeller Institute for Medical Research in 1910. The Rockefeller Foundation set up a suite of offices for its own people as well, presumably the better to carry out their fund-raising efforts on behalf of the MBL from 1919 to 1924. It was called the Rockefeller Lab until its demise.

The communal class laboratories have always provided the opportunity for students to work things out together. They can talk about the problems or the new ideas they are having, then get suggestions or try them out with colleagues working on similar research. This is the sort of opportunity for cross-fertilization that the MBL is all about. Late at night a visitor can find earnest prospective biologists at work, worrying through some problem or other and exploring how to get things to work next time. Such dedication has typified student behavior throughout the hundred years, although today's enrollee experiences far more distractions at the MBL and

Laboratory in Old Main around the turn of the century. Standing in the center is Leonard W. Williams, instructor. MBL Archives.

elsewhere in the modern world. Easier transportation makes it all too simple, for students today may forge out into the real world, whereas earlier decades found them comfortably confined to Woods Hole. As long-time MBL embryologist Cornelia Clapp wrote in 1888, it was even a bit dangerous to venture outdoors at night because of the large boulders left by the glaciers that had once covered the Cape. These cluttered the MBL land and kept people indoors, at work.

In those early days, most of the students in the courses were neophyte researchers. High school science teachers or college teachers without much research experience, they delighted in the introduction to diverse life forms and to serious science. More advanced students signed on as investigators and pursued their own work, directed by the senior staff. With time, the students in courses have become increasingly advanced, and those wide-eyed, eager beginners go elsewhere. Very few undergraduates populate the MBL courses, though the recent program to involve high school students in doing their own research at the MBL demonstrates that seventeen- and eighteen-year-olds are perfectly ready for scientific work and may even be more curious and more willing to try new and risky things

The general zoology lab in Old Main with Leukart charts and drawing on the blackboard.
MBL Archives.

William P. Procter standing with his bucket on the marsh of Great Harbor. MBL Archives.

William Procter at his lab table in Old Main, 1923. Photograph by Alfred F. Huettner, MBL Archives.

than some older students. Students have an enthusiasm and resilience which has benefitted the MBL throughout its 100 years.

In the early days of introductory courses, nearly everyone congregated in the central classrooms or around the lamps at night in the laboratory to draw and to record the observations of the day. If it was cold, they lit up bunsen burners to drive off the chill. The shared general laboratory for students made the close collaboration and exchange of ideas almost unavoidable. The initial wooden building was designed with lots of windows to let in the light for microscopic work, and was organized around a central site for seawater where organisms resided away from the disturbing sunlight.

This century has brought photography, photocopying, and round-the-clock lighting to allow microscopic observations at any time of day or night. One no longer has to worry much about exposing the embryos to too much sunlight, which might speed up or artificially alter development. The laboratory designs in the new building reflect the advances, because windows no longer have to function as the major source of light for microscopic work. But people do continue to make changes, such as adding aluminum foil to create light-controlled environments or air conditioning for midafternoon comfort.

The buildings now are designed with separate lab spaces for researchers. People from different parts of the country working on similar but different problems find themselves not shoulder to shoulder or through only an open door from others, but in the next lab. In the earliest years in

Old Main even people in the separate thin-walled rooms upstairs knew what the others were doing. Now researchers have more privacy, enough that they may never even see or meet each other. More people, with their special needs and interests, have brought changes.

The separation does have advantages in our increasingly specialized world, which demands continued output and progress of its students—not that those early leading researchers did not produce as much and as high quality work as any competitor today. Working together in one lab can bring inconveniences such as noise and lack of privacy. University of North Carolina embryologist Donald Costello reported that when he had worked at the MBL when M. H. Jacobs was director, he caused problems with the noise from his centrifuge. Jacobs felt that eight linear feet of counter space should be enough research space for anyone. Costello found himself in a room with twenty-six other researchers, though his own eight linear feet did not actually connect with anyone else's. His work required centrifuging, and the typical centrifuge of the day put out a terrific racket. Since he ran it for up to twelve hours a day, nonstop, the noise drove him a little crazy. He would go out to sit by the waterfront and recuperate, but not all of his labmates could do the same, as some were engrossed in projects that demanded their constant attention. Costello asked Jacobs to give him a separate place for at least his centrifuge in order to spare the others; but the director responded that no one should object to the sounds of science. Only investigators with significantly more research stature were entitled to private work space in those close quarters. In fact, considerable effort went into planning whose laboratory space would be where in the building.

When Old Main went down, reportedly because its old wooden structure no longer met fire codes, many people mourned its demise. "Can't we spray it with something?" supporters asked. With a grant obtained to construct a new building and inspectors' reports militating against remodeling, the old building fell. Pieces of its familiar shingles reside among family treasures in many Woods Hole cottages. Old Main was a place for everyone to know what everyone else was doing. It began in 1888 as one 63-by-28-foot two-story frame building, erected at a total cost of $1,600, on land costing $1,300 for a 78-by-120-foot lot. Whitman then added another section in 1890, and another in 1892. The total reportedly cost about $3,000 and, as Costello suggested, with that initial investment "probably more significant scientific work was done than has ever been done in any $3 *million* building ever created since." That building, one might reasonably claim, trained the first generation of American experimental biologists.

Other separate wooden buildings of the same unassuming style followed thereafter, with an alarming regularity that threatened the trustees' sense of economic well-being. Certainly, the MBL did continue to attract more and more people each year, but that might be only a temporary

Old Main in 1888. It was constructed for a total cost of $1,652 in just under three months' time. MBL Archives.

success. These trustees, who had made themselves legally and financially responsible for the Laboratory, worried about Whitman's enthusiasm for expansion. Only later did the Laboratory begin to acquire other gifts of land, including their first piece of waterfront property in 1902. After that, various gifts and purchases, most notably from Charles R. Crane, extended the MBL holdings and made expansion more secure.

Besides augmenting the MBL's operating budget, Crane also donated funds to the MBL to construct a new laboratory building, the Crane Building, which was erected in 1913–1914 as the first permanent (brick) MBL structure. In the dedication speech for that new building, Crane also proudly referred to the spirit of research that inspired him to want to contribute. "Without that spirit no amount of bricks and mortar and organization would be of any great service, but with that spirit the laboratory has been able to accomplish a great deal with very simple means." That spirit represented freedom and cooperation and democracy, Crane continued, and the time had come to give it "a more substantial body."

Crane and Lillie then accumulated further funds to add on another wing, creating an impressive L-shaped brick edifice. The new wing was called the Brick Laboratory for a while and only much later the Lillie Building. The Crane and Lillie complex, now often mistakenly simply called Lillie, was a fairly straightforward brick building: upright, solid, sturdy, reliable, with all the necessary conveniences. When they added a section to the library just before 1940, they had to scrounge around all over New England to get the required materials as the impending war had made things tight. They managed to complete the project just before World War II began. The Lillie and Crane buildings are more functional and restrained than fashionable or ostentatious. They are solid and have served their purposes well and promise to continue doing so, which is what really matters.

Like Lillie himself, the Lillie Building now serves as the rather austere and quiet control center for the MBL. The mail room is located in the basement, as are the chemical supply room, the equipment room, and other, unheralded functional rooms and their staffs that make the MBL run. Laboratory equipment arrives at the receiving room and thus begins its MBL life in Lillie. The administrative offices used to reside in Lillie until 1981, down the hall amid the clutter of laboratories and other offices. Recent remodeling of the Candle House has provided a more appropriate, fancier, and more modern setting for the administration, now removed from the bustle of everyday science.

Lillie also houses the large auditorium, where the MBL community congregates for the famous Friday evening lectures. In the early days, when Whitman was impressing his vision of the MBL on the place, the evening lectures served a somewhat different purpose than they do now. Then they were directed at problems. Rather than reporting research results, they were designed to discuss ideas of research approaches more generally. Because a high percentage of the community shared overlapping research interests, the lectures could appeal to nearly the entire audience. The modern listener shares the silent presence of many of the MBL greats, whose names are inscribed on the refurbished mahogany chairs in the auditorium. One can gaze at Crane's portrait on the wall, marveling that he really does not look like such an imposing person, even though he did silently and single-handedly save the MBL during its worst economic years.

Outside the Lillie Auditorium, around the corner, sits a bronze statue of Confucius. Mr. Crane had served as minister to China in 1920 and brought back Confucius for the MBL. Originally placed in the Crane wing of Lillie, Confucius has wandered, at times to the MBL beach and even to Nobska. For a while he sat watch near the watchman's desk, and now he has returned to his original location. Tradition holds that a researcher who places a penny in Confucius's hands will have rewarding research results,

The itinerant statue of Confucius in the lobby of Crane. MBL Archives.

whereas those who fail to observe the custom will only publish in the *Journal of Negative Results*. Children in the know regularly check to see if there are any coins in place, and researchers caught putting pennies in explain that they are only "for the children."

Crane Building and Candle House in 1923. Photograph by Alfred F. Huettner, MBL Archives.

Funding and Control

The original trustees tended to have a conservative vision of what was possible or expedient. The group included Hyatt, of course, and several other professors of zoology or botany in Boston (William Farlow, Charles Minot, and William Sedgwick), but also representatives of the Boston Society of Natural History's ancestral links to the MBL (Florence Cushing, Susan Minns, Samuel Wells). At first, this mixture of scientific and nonscientific individuals enthusiastically supported the MBL projects. They raised money by giving lecture series and sponsoring concerts and other popular events of the time, and they circulated announcements of the Laboratory session. Whitman helped with much of the fund-raising as well and gave his own time, potential salary, and additional monies to help the MBL.

Every year, Whitman presented his expanding plans to the trustees. Sometimes he just went ahead and made commitments and spent money without official permission. Some of the trustees began to rebel. They fretted about the increasingly professional and research-oriented direction of the MBL and about what some saw as Whitman's desire for increasing control, as well as his tendency not to worry very seriously about running at a deficit. At the same time, the number of trustees gradually grew larger. Some were added because of their scientific reputations; others, because they were important people and held the prospects of providing financial support, or because they had particularly endorsed the basic idea behind the MBL summer school. The balance began to shift away from the natural history school plans of the original group and increasingly toward the more expansive and less locally oriented idea of developing a leading research laboratory.

Major changes came about in 1897, following a minor revolution. After the 1896 session some of the trustees concluded that Whitman had finally overstepped all reasonable bounds in spending funds without official permission. They refused to provide further funds and even threatened to close the Laboratory altogether. Eventually, the crisis was resolved at least long enough for the summer session to continue more or less as planned. Fewer students did attend, however, because the trustees' disagreements had delayed the announcements, and people simply did not know whether the MBL would even exist in 1897. School teachers had to make plans and often had only limited financial support to attend such a school; they could not afford to wait until the very last minute. This disagreeable state of affairs simply could not persist. So the scientists, with Whitman at the helm, staged a quiet revolution.

At the corporation meeting in Woods Hole in 1897, the group voted for changes. The MBL Corporation has always played a role in directing the laboratory and technically owns the facility, but that role received consid-

MBL from across Eel Pond, early 1930s. The large brick building to the right is comprised of Crane and Lillie; immediately to the left is the supply building, with Candle House behind. Photograph by Alfred F. Huettner, MBL Archives.

erable strengthening with the decisions of 1897. Previously, the corporation had included the official incorporators, consisting of a few scientists and those representing the Boston Society of Natural History and the Woman's Education Association, of course, plus voting participation by "All who aid in [the laboratory's] support by subscribing to investigator's tables." This included a number of subscribing schools and societies. By including the subscribers in the corporation rather than focusing on those who actually occupied the tables, that governing group was weighted toward administrative and institutional representation.

With the troubles of 1897, Whitman and his supporters, such as the influential Columbia University and American Museum of Natural History paleontologist Henry Fairfield Osborn, decided that the scientists should exercise greater control. They particularly distrusted the strong Boston representation on the board of trustees. Though Whitman consistently envisioned the MBL as a *national* facility, as he stressed over and over again, some trustees thought he really meant to take control himself.

This conviction about the principle of control by the scientists for the scientists, while widely praised in the years since, probably had less impor-

The Chem Room, 1952. Photograph by Alicia Hills, MBL Archives.

Candle House in the early 1920s before Lillie was added. It has served the MBL as a supply area, dormitory, administration headquarters, and lecture/meeting space. MBL Archives.

The supply department when it was in Candle House. MBL Archives.

tance in the 1897 reform than Whitman's eagerness to rid himself of those hopeless Bostonian "old maids." He was not at all opposed to women as scientists. Indeed, his own wife, Emily Nunn Whitman, had studied as a biologist in Europe and the United States and had taught at Wellesley College. But those Boston women trustees were not scientists, and he felt that they failed to understand what would make biology work. In short, they did not share Whitman's vision for a national biological station that would embrace advanced research as much as it did teaching.

In 1896 and 1897, Whitman and his cohorts plotted change. They set up a reorganization committee. They officially appointed L. L. Nunn of Telluride, Colorado (Whitman's brother-in-law), as a trustee. Mr. Nunn then made a generous offer to provide financial security for the MBL's deficit for that session. But, he insisted that

> I make this offer upon the condition that your present director, Dr. C. O. Whitman, directs the affairs of said Institution in accordance with his best judgment during the said season, and that Mr. L. Wilcott Allen, or

> some other person acceptable to me, be employed as Assistant-Treasurer at Woods Hole, with full authority to collect and receive all revenues and income from said operations, and to supervise and direct the expenses incurred in said operations, with the understanding that I am not to be responsible for any expenses, except those incurred with and by the knowledge and consent of said Assistant-Treasurer or of myself.

Nunn had already placed $500 in the hands of Mr. Allen to guarantee the opening of the lab. Mr. Nunn's hard-hitting, straightforward western style appropriate to silver mining in Colorado did not sit well with all of the proper Boston trustees. They wanted to retain control themselves. On April 12, 1897, they voted to open the MBL for the summer but to reject Nunn's offer. It seemed, after all, that they could find enough money for the purpose without giving up control to Whitman or to any such outsider as Nunn.

The trustees as a group wanted to retain control of their laboratory, for which they had initially accepted legal and financial responsibility. In contrast, Nunn endorsed Whitman's desire for more leeway as director and for not always having to have everything approved by the trustees.

On August 6, the trustees met once again in Woods Hole. After considerable worry about financial details, they voted to add new members to the corporation. One hundred and fifty new members joined the list. These people included mostly investigators and students at the MBL and significantly swelled the ranks of scientists in the governing body. Then 4:20 P.M. came closer. The Boston trustees had to leave to catch the last train of the day if they were to avoid an expensive overnight stay. This left the balance of scientists, who lived in Woods Hole for the summer. Of the nonscientific Boston group, only the secretary, Miss Anna Phillips Williams, stayed. She found herself in a sudden and shocking minority. The real changes began in what must have been a positive and emotionally charged meeting.

First, the new majority voted to pay for salaries in priority to other bills—a shift of the position held by the trustees as a whole. Then they voted to move the annual meeting from Boston to Woods Hole during the summer—over Anna Phillips Williams's objection. They called a special meeting of the new corporation to change the bylaws, to be held in ten days—against Anna Williams's lonely objection. The group then voted to appoint a committee to revise the bylaws—over Miss Williams's continued objection. She was not appointed to the revision committee.

The crisis had passed. Most of the Boston trustees quit. And at the special meeting the new corporation changed the bylaws so that the group of trustees had to become more responsive to the scientific community.

Whitman and his supporters had won, so that the MBL became an organism on its own, in their view not so dependent on the action of that nucleus of restrictive conservatism represented by the nonscientific Boston trustees. With the help of Mr. Nunn's money, the MBL could continue to grow and flourish in the manner to which Whitman wanted it to become accustomed. Building and expansion continued apace, and all was well—briefly.

But running a "national" laboratory proved horrifically expensive. What the MBL, like any aspiring institution, really needed—and still needs to this day—was a solid major endowment. For a while it appeared that Helen Culver, who controlled the Hull fortune in Chicago, which had also supported Jane Addams's Hull House project, had provided just such an endowment. She had donated one million dollars to the University of Chicago, with the intention that it also be used to provide classes for Chicago's less well-to-do West Side and to support a national marine laboratory, among other biological activities. Her correspondence makes her intentions clear. She hoped to support the MBL as well as the University of Chicago because advisors had convinced her of the value of both. As plans unfolded for the projects at Chicago, however, it became evident that her money would not stretch as far as she and her advisors had hoped. Her donation came in the form of real estate, and prices had dropped pathetically in a short time. In addition, building prices had risen so much that providing a modern laboratory for the University of Chicago finally took all the million dollars, as well as an additional gift from Miss Culver. The MBL gained nothing but moral support from the Hull fortune.

By 1900 the continued difficulties in raising money, the demands to expand the laboratory, and the desire to include more advanced (and more expensive) investigations combined to make Whitman even more anxious to obtain financial security. L. L. Nunn and a group of businessmen representing the University of Chicago made an offer that it seemed unlikely the MBL could afford to refuse. The group offered a considerable sum of money in exchange for some simple expectations of financial accountability to them. The offer seemed very attractive. Stubbornly and perversely it seemed to Whitman, the MBL board nonetheless refused. Whitman felt that some of the trustees had impugned his integrity with their implication that he and Chicago were trying to take over control of the Laboratory. It seems clear that this charge was unfair to Whitman, who had after all served as director all those years with no salary and often at some considerable personal expense. Whitman wanted to place the MBL on solid financial ground. But the Nunn proposal was unacceptable to the MBL's leading and trusted scientists.

Whitman, in turn, opposed an attractive proposal by the Carnegie Institution in 1902 to fund the research of the laboratory. The Carnegie Institution by this time was beginning to fund various sorts of biological projects. Supporting the MBL seemed within their purview. But Whitman and others did not want to give up the independence of the laboratory and felt that Carnegie money would mean Carnegie control. The MBL trustees did not accept the Carnegie offer, though they did receive three years' worth of support in meeting expenses from the Carnegie people.

Whitman retired as MBL director in 1908, in part because it was too much trouble to haul the pigeons that he had begun to study back and forth from Chicago to the Cape each summer, in part because there was no longer any pressing reason to go to Woods Hole, and in part because he had tired of all the battles. His personally trained assistant, F. R. Lillie, took over. Whitman had directed Lillie's dissertation, which was begun at Clark University and completed at the University of Chicago. Then he had exerted considerable effort to hire Lillie at Chicago. Though not successful at first, he eventually managed to attract Lillie to Chicago from Vassar. Lillie became assistant director at the MBL and assistant chair at Chicago, in many ways simply taking over control when Whitman no longer wished to have it.

Lillie had money, and Lillie attracted money to the MBL. In fact, the MBL owes much to Lillie and to the new booming business of indoor plumbing. Lillie's brother-in-law was Charles R. Crane, a wealthy businessman associated with plumbing and porcelain fixtures. From 1904 to 1923, Crane helped to make up the MBL's deficit, amounting to roughly $20,000 per year in the 1920s.

In fact, the MBL began really only in the 1920s to attract any major endowment and financial assistance, well after Whitman's death in 1910. During that period, it gained sufficient support both to enlarge the facilities and offerings and to secure what was then a substantial endowment. The National Research Council (with many MBL researchers as members) supported work at the MBL and called for financial investment by others. The Rockefeller Foundation, John D. Rockefeller, Jr. personally, and the Carnegie Institution all made liberal contributions, with various conditions for support to be provided by the MBL itself. That indispensable benefactor Charles R. Crane met these conditions brilliantly and thus put the MBL on a "high plateau of security."

The successes of the 1920s took the MBL from one original wooden building to a physical plant worth $1.5 million, with an endowment of over $1 million. Even that magnificent sum proved insufficient to do much more than desperately tread water during the Depression, when annual income fell and attendance dropped drastically from 1931 to 1935, but it did keep the MBL afloat. Then came the great period of government funding (but not government control) with the strengthening of the National Research Council, then the advent of the National Science Foundation and National Institutes of Health to replace the private Crane, Carnegie, and Rockefeller support.

A painting of benefactor Charles R. Crane, which hangs in Lillie Auditorium. MBL Archives.

During the building years of the 1920s, Crane wrote to Rockefeller to express his pleasure at being involved and his satisfaction with the results, emphasizing the important spirit of the place: "These scientists were struggling and accomplishing marvelous things with most meager equipment, making many sacrifices—It seems to me that the precious thing to preserve was the spirit of the organization, a spirit everywhere recognized although hard to seize or to imitate." His generosity and the great successes with fund-raising helped to stimulate the General Education Board to add its contribution, giving substantial funds to help the library purchase back series of journals and to provide an endowment. They also gave another $250,000 to support the dormitory and apartment house projects, which provided the first MBL housing facilities with winter heating and with housekeeping facilities for families. The 1920s proved an active time of strengthening finances and resources at the MBL as in much of the rest of the United States. Further financial relief came only well after World War II, as building grants have allowed the MBL to expand physically and to gain security.

NOTES

Donald Costello, in a transcript of a taped interview at the University of North Carolina, 1967, mentions the importance of such factors as temperature for isotope use.

Robert Kahler's interview, Historical Collection, gives an eyewitness account of the various fires, which are referred to by a number of other interviewees as well. Isabel Morgan Mountain, in her valuable unpublished history of the Morgan barn property and in a 1987 interview, places the fire in historical context.

Winterton C. Curtis recalls his being drawn to the MBL in "Good Old Summer Times and the M.B.L." and "Rhymes of the Woods Hole Shores," *Falmouth Enterprise* (August 12, 19, 26 and September 2, 9, 1955), reprint in MBL Archives.

Laboratory allocations and the desire to change the system are discussed in MBL annual reports and other archival notes. Whitman often discussed his views on education in MBL annual reports and in a number of documents and letters both in the MBL archives and at the University of Chicago Archives, as well as in lectures in the *Biological Lectures* of the MBL.

Philip Pauly's excellent *Controlling Life* (New York: Oxford University Press, 1987) provides a provocative and important account of Jacques Loeb's work and of his role at the MBL. Also see Pauly on the MBL's place in history, "Summer Resort and Scientific Discipline: Woods Hole and the Structure of American Biology, 1882–1925," in Ronald Rainger, et al., editors, *The American Development of Biology* (Philadelphia: University of Pennsylvania Press, 1988), pp. 121–150.

Other Loeb stories, by W. J. V. Osterhout and by others, are on file in the MBL Archives.

Donald Costello discusses his early times at the MBL in his 1967 interview, deposited in the MBL Archives.

The initial building bill resides in the Archives, along with other documents about costs.

On the 1897 and other crises, see Lillie's history (Notes Chapter 1), MBL trustees' minutes and annual reports, and papers in the *Biological Bulletin* Supplement. See also "A Statement Concerning the Marine Biological Laboratory at Wood's Holl, Mass.," *Science* (1897) 6: 529–534 from disgruntled trustees and "A Reply to the Statement of the Former Trustees of the Marine Biological Laboratory," October 8, 1897, as a separate publication because they did not feel that they should air MBL business in public. See also Conklin, "The Reorganization of 1897," *Collecting Net* (August 17, 1935): 209–211; and Jane Maienschein, "Early Struggles at the Marine Biological Laboratory over Mission and Money," *Biological Bulletin* (1985) 168: 192–196.

On later support from Charles Crane and others, see MBL annual reports and *Collecting Net* throughout the 1910s and 1920s. Also a letter from Crane to J. D. Rockefeller, Jr., December 22, 1923, records his affection and respect for the spirit of the MBL.

4
The Library and Publications

Across the street from the library in the Lillie Building. Photograph by Hugo Poisson. MBL Archives.

the Collecting Net
The Marine Biological Laboratory
Vol. 2, No. 5
August 1984
Grass Fellows
Do It All
See Page 8

A MONDAY MORNING TOUR through the MBL campus reveals one area of especially intense hustle and bustle. Carts of books, with their request slips indicating desired page numbers to be copied, line the halls outside the copy room. Piles of volumes to be reshelved cover the countertops. MBL scientists and students have busily put the library to use over another weekend, as always. Even on a stormy weekend night, with the wind blowing wildly and setting up an unsettling moan throughout the building, people work quietly in various tucked-away corners of the library stacks. Today the quiet clicks of word processors at all hours augment the sounds of turning pages.

The Library

In many ways the real center of the MBL lies in the library, run since 1962 by head librarian Jane Fessenden. No locked doors or guards or stern looks interrupt the researcher who comes to work in the library. Instead, library readers meet friendly faces and cheerful smiles. Some people think the library staff is not serious enough because they seem to have too much fun. They are always there, ready to help with even the smallest questions. They also make it clear that they believe a library is there to be used, and not primarily to be preserved for some unsuspecting and very possibly unappreciative posterity. Anyone with a legitimate purpose, who gets prior permission and who follows the rules, may use the MBL Library at any season of the year and at any time of day.

At first, the head librarian was a scientist who volunteered to take care of the books. Then the lab hired an assistant, and then a full-time librarian who would listen to the scientists and set up the operation the way they wanted it, someone who would want to go to them for advice about their library.

The scientists respect this valuable resource, as they always have. Some take it for granted: of course they can run in, find what they want on the shelf, have it copied, and get back to work quickly and easily. Some of these

The main reading room, as it looked in the fifties. Recently refurbished, it is now called the Bay Reading Room.
MBL Archives.

researchers might even claim that the MBL Library is a useful convenience but hardly the real core of the research center. Yet they too publish their results, in journals and books that become part of the long-term scientific world housed in the library.

Each year scores of scientific researchers recognize the value of that resource. They come to the MBL as library readers to write books, revise textbooks, or complete long reviews. One recent year found more than 150 readers from many states and countries working on a wide range of historical and biological projects. People go to great effort to spend a sabbatical year in the MBL Library, where more than 400 years' worth of biological literature is kept. As Harvard paleontologist Stephen Jay Gould has said, "The library at the MBL is an institution that has its own humanity and seems to me more an organism than a place."

As such, it has been the recipient of practical jokes. In 1926, at the height of the first major American bout with creationism in the wake of the Scopes trial, the editors inserted the following item into the MBL's newspaper, the *Collecting Net*:

> We were shocked beyond measure last Sunday morning when an exhaustive search failed to reveal that classic book on evolution—the Holy Bible—in what is supposed to be one of the finest libraries of its kind in the world. The situation was untenable. It could not be allowed to stand.
>
> Fortunately the Trustees of the laboratory saw fit to severely censor [sic] the Librarian at a special meeting on Tuesday afternoon called for that purpose. The Editorial Staff wish to commend the Trustees for their prompt and efficient action in relieving the situation.

Creationism attracted attention again a few years ago with a lecture series on science and religion and a recent speaker who asserted that no honest scientist can possibly be religious. And numerous books about science and religion continue to appear on the library shelves.

The first year, the MBL Library began as a handful of books in the back corner of the downstairs general laboratory room. By the second year, a healthy collection of 343 volumes occupied the library corner. Over the next years, Cornelia Clapp served as librarian for the slowly expanding collection that moved from corner to corner in search of sufficient space. During the crisis of 1897, the trustees ordered Whitman to stop spending money. Notwithstanding the injunction, he asked Clapp to continue ordering journals. The trustees then informed her that she could pay for them all herself. With that in mind, Clapp nonetheless did order the volumes, which were, in fact, paid for. Yet the uncertainty illustrates the tenuous basis of the MBL in its earliest years and the scientists' commitment to maintaining that library as a vital organ essential to the life of the MBL.

In 1914 the MBL hired an official assistant to the librarian for the first time, when the library moved into more substantial, fire-protected sur-

The library stacks when they were in Crane 217 before Lillie was built in 1924. MBL Archives.

Cornelia Clapp, early librarian, shown here circa 1933 with the board of trustees, of which she was a longstanding member.

roundings in the new Crane Building. They used a wheelbarrow to haul all the books across the street. For the first time someone was actually paid to run the library. In 1919 Priscilla Montgomery took the assistantship role. She became librarian in 1925 and continued in that position for many years. The widowed wife of University of Pennsylvania cytologist T. H. Montgomery, she was part of the MBL community and understood what the scientists wanted.

During that time, the library and its holdings continued to expand, even during World War II, and moved into larger quarters in the new Lillie Building when it opened. A section of that building was specially designed for stack floors. There the library remains, now complete with up-to-date computerized interlibrary loan and other searching services. Further expansions and additions have continued to help the facilities keep pace with demand, more or less.

Open windows bring in the fog, which threatens the books at times, but the staff keeps the books and bound periodicals in order, properly on the shelves, and in good condition so that the volumes are often in far better shape than those in a fancy, environmentally controlled setting where they are never used. Leather-bound books like to be handled, and respectful use over the century has caused little damage beyond normal wear. The greatest miracle is that the system works. Loss due to theft is lower than in almost any other institutional library in the country. The absent-minded professor may wander down the hall to jot down a reference and intend to bring the volume right back. The book is therefore temporarily lost and unavailable to other readers, but such books almost always find their way back to the library—eventually.

Sometimes books return in an unorthodox way, as they did one time during World War II. University of Michigan cytologist and botanist William Randolf Taylor occupied an upstairs room of one wing of Old Main, the room at one connecting corner in the U-shaped building. The false ceiling had openings in it, which raised the curiosity of the small number of

A quiet place to work: one of the sixty-eight desks located in the MBL Library stacks.
MBL Archives.

students still in attendance during the war. The students wanted to know what was inside the false ceiling. Taylor responded that there were probably only dead birds, which aroused the curiosity all the more. So the students pulled up benches and crawled in for a look. They found books. Bushels of books. Some had long since been replaced when they were discovered as missing, but here they were, bird-stained and rain-beaten. Nobody recorded what happened to the unexpected treasure.

What is more amazing, however, is that the journals remain in their proper places. The staff checks carefully to ensure that volumes are not misshelved. The researcher can generally find nearly all of the material sought, though the journals sit on the shelves in alphabetical order by title, which sometimes sends the reader to each of five different stack levels in search of any particular subject. Because of the tradition of gathering books and continuing journal subscriptions even during tough economic times, the MBL Library has full runs of a large number of rare journals. Roughly 80 percent of the journal titles go back to the first volume. In 1924 and 1925, the General Education Board's gift of $50,000 provided sufficient funds to purchase back sets of incomplete runs, which filled out the collection. In addition, the board provided an endowment for library use.

A recent $2.5 million matching grant from the A. W. Mellon Foundation provides a further endowment for the library and puts its operation on a

The card catalogue when it was in the main reading room. MBL Archives.

Recent periodicals, kept available in the main (Bay) reading room until there are enough to be bound, a process that takes only two weeks. They are then put in the stacks. MBL Archives.

more secure financial basis than it had been on. These endowments are appreciated, for as demand has increased, so have costs. When the Woods Hole Oceanographic Institution was founded in 1930, the two institutions agreed that having two separate, overlapping libraries would be foolish. The MBL would run the MBL-WHOI Library, as it has been called since that time, because the MBL already had one, and WHOI would buy books and journals and contribute to operating costs. In those beginning years, there was not much of a science of oceanography, so the addition did not require a host of new journals and books. But time has changed that, so that during the last decade expansion has also dictated a more formal arrangement between the two institutions.

In addition, the MBL is now the official library for the National Marine Fisheries Service, originally the Fish Commission. In the early years, the Commission men wandered over to look through the MBL's book corner. They then set up their own facility. Only recently did the U.S. government put the library out for bid, at which point the MBL was officially given the contract. In addition, the library belongs to the Southeastern Massachusetts

Consortium of Hospital Libraries. This gives same-day access to a wide range of medical journals that the MBL would not otherwise have.

In 1983, the library undertook a use survey of the journals, with financial assistance from the Rockefeller Foundation. For ten months, readers recorded every time they used a journal, and the staff compiled the results—for every bound volume and current periodical in the library, not just some selected samples. The survey showed that the full complement of holdings are used actively, even seemingly obscure, very specialized, and old volumes. Furthermore, even those volumes used relatively rarely prove valuable in special circumstances. Scientists find the completeness of the collection, with its current holdings of 160,000 journal volumes, a resource matched by very few other places.

One problem that demanded a solution by the mid-1970s was that of space. All those different runs of journals, each slowly increasing in size, take ever more space. Careful evaluation and management provide predictions for expansion. With the help of generous grants, the MBL recently undertook renovation and remodeling of the Lillie Building, including the library. Eleven laboratories were removed to make room for the library expansion, and a modern climate-controlled rare books room and archives was built. Everything had to be moved—by hand. And everything found an adequate new home except the old reprint collection. Not enough space, they said. The once extremely valued and fully catalogued collection of reprints lost its appeal as the dust gathered and as copy machines made private collections easier to acquire. History-conscious researchers miss this wonderful unique resource of more than 250,000 reprints, now largely languishing away in a storage hallway, but most of the materials are available in the original journals. And eventually, the librarians insist, the collection will find a new space.

Publications

Since 1908, when Lillie effectively took over as director of the laboratory, the MBL has been following Baird's example at the Smithsonian and has exchanged copies of its journal, the *Biological Bulletin*, for other journals. In the 1940s, the library acquired roughly half of its 1,200 titles that way. Now the exchange accounts for about 600 of the 3,000 periodical titles currently received.

The *Biological Bulletin* began after Whitman decided to begin a new American publication. This happened before the MBL began, when Whitman was director of the Allis Lake Laboratory near Milwaukee, Wisconsin. A businessman from a business family, Edward Phelps Allis had wanted to do biological work. Advice from England and America led him to set up his own laboratory and to hire a director/teacher. When Whitman took that part, he suggested to Allis that an American journal would prove extremely valuable, and that it was a disgrace that none existed. Allis promised support, and Whitman began editing the new *Journal of Morphology*. Designed to carry long papers with what Whitman regarded as the requisite degree of detail and quality, the journal proved very expensive to run. Conklin's dissertation in 1897 cost $2,000, and others must have come close. Whitman never settled for less than the best possible, however, so the journal continued publishing excellent and finely illustrated articles until it had to close down briefly because of the deficit. Fortunately, the Wistar Institute resumed publication soon after.

In addition to the *Journal of Morphology*, Whitman soon recognized the advantage of having a companion publication for shorter articles that could be published quickly. This took the form of the *Zoological Bulletin*. After two volumes in 1897 and 1898, however, Whitman and his cofounder, Harvard entymologist William Morton Wheeler, decided to transfer editorship to the MBL and to change the title to the *Biological Bulletin*. Its three-dollar price for three hundred pages produced a per-page cost that remained about the same into the 1940s, after which it joined the escalating world market. It has always been intended as a general journal, never restricted to MBL research or researchers (only 9 percent of the articles in the most recent year were by MBL scientists), and includes work on a variety of biological topics. More recently, it has published papers from special symposia and the set of abstracts of papers presented at the General Scientific Meetings held each August at the MBL.

With these two publications in his power and as leader of the MBL, Whitman became a very influential figure in American biology. In addition to these periodicals, the MBL also issued *Biological Lectures* in order to publish the Friday evening lectures under his directorships. With time, of course, numerous other journals have appeared, including a considerable number begun by MBL researchers.

One such periodical was the *Collecting Net*, begun by Ware Cattell. Cattell was an unusual individual who specialized in electrophysiology and inspired various enterprises, such as this popular newspaper. The paper was put together each week by Dot Rogers and a friend on their way over to the printer in New Bedford. They worked on the boat ride over, delivered the copy, waited for the printed papers, and returned with them in a short time. The more recent version of the *Collecting Net* is put out by the Public Information Department, under the direction of George Liles, and serves to inform the wider MBL community of activities there. The first run of those wonderful mixtures of science and gossip resides in the Rare Books Room and Archives of the library.

Rare Books Room and Archives

In fact, the Rare Books Room is one special and little-known feature of the library. Accoutered with Louis Agassiz's intriguing leather-covered swiveling table, this comfortable area houses many marvelous collections of photographs and archival materials as well as the extremely valuable rare books and journals, many of them gifts from appreciative MBL alumni. One recent addition came from T. H. Montgomery's sons and includes volumes detailing Captain Cook's voyages. Few students find their way to this room, but scholars wanting to check out the history of a subject or to find illustrations of the MBL or of work done here can make arrangements with the archival staff to use the materials.

In 1986 WGBH, one of Boston's public broadcasting stations, sent a team of photographers and writers to do a special show on the MBL. They also chose the MBL to photograph for a different set of shows a number of rare prints and book illustrations dating back to 1560. When asked why the MBL, especially when Boston has its own libraries with some of the same rare volumes, the crew replied that the MBL staff was so friendly and helpful. Some of the Boston facilities wanted to close off their materials rather than to make them available to the public even through film. By contrast, the MBL Library's attitude of respectful sharing for purposes of scientific research is historically typical of the MBL.

One day not long ago, three topnotch scientists sat at Agassiz's desk in the reading room of the Rare Books room. Photographs of former great researchers hung on the walls. One of the scientists was redoing his cell biology textbook and expanding the historical parts of the introductory section by rereading the classic texts that reside in the Rare Books Room. Another knew that work very much like what he was doing in his lab had been done at the MBL in the 1890s by E. G. Conklin, and he was checking the earlier papers in the bound reprint collections. The third was a visiting historian looking for a photograph to illustrate an upcoming lecture in Paris. At the same desk sat a first-rate science writer, pursuing his own study of global ecology, quietly reading books by Agassiz, Alexander von Humboldt, and other nineteenth-century writers who worried about the history and dynamic process of the earth as a whole. The researchers spoke briefly, curiously looked over each others' shoulders for a moment and exchanged ideas, admired the rich written and photographic resources around them, and then settled in for another typically intense day of research at the MBL.

NOTES

Jane Fessenden has helped clarify many episodes and provided many leads for discussing the history of the library. The MBL annual reports have a library section which provides information for each year; they also document the statistics of journal use and loss rates.

Stephen Jay Gould is quoted in a library flyer, n.d.

The *Collecting Net* story about the Bible appeared August 26, 1926, p. 5.

William Randolf Taylor's interview, Historical Collection, discusses the finding of books in the ceiling. On Whitman and journal publication, see Ernest J. Dornfeld, "The Allis Lake Laboratory," *Marquette Medical Review* (1956) 21: 115–144. See also Pamela Clapp, "The History of the *Biological Bulletin*," *Biological Bulletin* (1988) 174: 1–3.

Dot Rogers told us in a 1987 interview about riding the boat over to get the *Collecting Net* printed each week.

Details and data come from the MBL Library's official records, MBL Archives.

5
The People

Collecting party at Cuttyhunk Island, 1895. Captain Veeder (on the right, with a child on his shoulders) had served the MBL for many years and had grown up at Cuttyhunk. Photograph by Baldwin Coolidge, MBL Archives.

Collecting isn't all work aboard the Dolphin *(1952).*
Photograph by Alicia Hills, MBL Archives.

One little boy, about five, happily asked his father at a recent MBL Day after they had toured labs and looked around, "Dad, can I come here to school when I grow up?" That is the point of the MBL: biology at the seashore is attractive to a wide range of people.

Students

Political and economic changes may dictate which people are able to attend MBL sessions and which accents will predominate at the lunch tables in any given year, but the MBL tradition persists. Students come and mingle and learn, and then return as teachers to the familiar surroundings to do research with their own students in tow, or they send their students to take courses on their own. The MBL perpetuates itself that way, as new generations learn about the laboratory from its people.

Donald Costello, for example, reported that he wanted to attend the protozoology course at the MBL after his sophomore year in college. He applied and received a rejection. Embryologist Lester Barth advised him to go anyway and offered to give him a ride to Woods Hole. Barth thought that Lewis Victor Heilbrunn, with whom Barth was working, would probably set Costello up to work too if only Costello could afford the $50 required minimum fee. Because Costello had worked for the previous two summers as an office boy in a die manufacturing firm to earn money for his education, at forty cents per hour for forty-five hours per week, he had accumulated enough to embark on this zoological adventure. Once in Woods Hole, he discovered that he could sit in on all the lectures at no cost and could absorb the atmosphere of, as well as the ideas in, the place. This atmosphere kept him coming back for most of his summers, sometimes bringing students of his own and even instructing in the embryology course for about ten years.

Other students arrive in more conventional ways, applying and becoming accepted to their first courses. Often today they are far more advanced

By the 1950s, specimens were collected primarily by the supply department, but some classes continued to make field trips to study natural habitats for themselves. MBL Archives.

Mary R. Huettner, wife of the scientist-photographer, collecting at Sippiwissett Marsh. Photograph by Alfred F. Huettner, MBL Archives.

Embryology class, 1897, including Gertrude Stein, front left. Photograph courtesy of WHHC Archives.

in their careers—and far more specialized and sophisticated in their demands—than they used to be. Rarely do high school teachers seeking their first experience with living organisms populate the courses, as they once did. Nor do many people arrive so casually as Frank Lillie did when Whitman invited him to come on down to Woods Hole for the summer to begin his graduate career. No longer are there as many eager novices to be enthralled by their first invertebrate collecting trips to Tarpaulin Cove or Sippewissett Salt Marsh.

Most of today's neurobiology students, for example, are much more advanced and dedicated to their sophisticated coursework. They simply do not have time to experience what it is like to wade one's way through the meandering streams of a salt marsh and to become mired down in mud so that other students have to help pull you and your rubber waders out with a loud sucking smack. A few do head out on a Sunday morning to catch bluefish. Yet only rarely do the busy students have much free time to explore the intertidal areas nearby and to gain respect for the fragile ecological balance, though they may find time to play the time-honored pranks of attaching a crab's claw to someone's ear or a lobster to a long skirt (or to a pair of shorts today). Many still attend courses on their way to fame in other fields, as Gertrude Stein did while she was briefly a medical student at Johns Hopkins in the 1890s. Overall, a more advanced group of dedicated researchers come today.

The late night gathering around the bunsen burners to cook clams, lobsters, crabs, and occasionally even the ubiquitous snails is more likely to begin with a trip to the fish market or the supply department than to the beach, as it used to be. Strict regulation of marine fauna as well as laboratory specialization has dictated that. But that regulation has been stimulated by ecological and embryological research carried out at the MBL and other marine laboratories. Times change. Details change, but students do still come to the MBL for their courses with the same eagerness that they always have. If there is sometimes less lighthearted fun and more dedication in their work here, that reflects the changes of science and of society as well.

Administrators

To the students involved today in their own research and coursework, it is not always obvious who the leaders are. The MBL has a scientific director who organizes courses and laboratories, but that is no longer sufficient. Now a full team of administrators has filled in, doing jobs that either did not exist in earlier times or that the director himself carried out. Conklin once remarked that his role as chairman of the biology department at Princeton for decades was not particularly difficult; he just borrowed a secretary one afternoon a week and polished off all the business. Whitman never even had a secretary at the MBL, much to the regret of those who had to read his rather difficult handwriting and his undated letters. In those days the trustees had to approve all financial decisions, even to spend as little as $1.98 for postage or $3.04 for alcohol (and much administrative effort seems to have gone into acquiring alcohol until World War II, not so readily available in pure form then as it now is). Today, we have a group of business managers of various sorts to run the laboratory. Some are required to guarantee conformity to legal requirements regulating animal care, scuba activity, liability coverages, and such. Others keep up with the expanding physical plant, increasing course demands, sophisticated equipment needs, or complicated financial arrangements. This complexity makes it harder to tell just who is in charge, though the scientists, in the form of a corporation, still officially hold control.

For a long period until he retired in 1986, Homer Smith served as business manager of the MBL. As the only major year-round employee, he must have appeared to the public as "Mr. MBL." In fact, one year when photographers came to take the necessary set of pictures to portray the activity of the laboratory, they came in midwinter. They found few scientists at work in their labs, surrounded by the fancy equipment the photographers sought. Only Homer Smith was at work. He was persuaded to play all the roles of the MBL in addition to his own. The result appears as a series of Mr. Smith shots: in the laboratory doing research, at the electron microscope making observations, reading diligently in the library, and so forth. The prototypical dedicated scientist at work, in other words. Did anyone notice that it was the same indispensable Homer Smith each time? The MBL spirit normally resides in the population of researchers, students, and administrators working together rather than in any one man.

Edwin Grant Conklin and Charles Otis Whitman

In earlier years, when there were fewer people, the leaders stood out more clearly. Whitman—solemn, dedicated, with his shock of white hair and upright stature—inspired trust. He was clearly a leader, those earliest students recognized. Edwin Grant Conklin recalled his first encounter with the already well known man. Conklin was a student at Johns Hopkins under William Keith Brooks, and Brooks continued to prefer the Fish Commission to the MBL. As a result, Conklin found himself in Woods Hole at the Fish Commission table doing research in 1891. Brooks had sent him up to the seashore to begin his embryological research and had advised him, when Conklin asked, to study the siphonophores. Upon his arrival in Woods Hole, Conklin asked where he could find some specimens. Nowhere, came the reply. Woods Hole had no siphonophores. Thus thrown onto his own resources, and without a telephone to ask instantaneously for further guidance from the home advisor, Conklin turned to the plentiful and efficiently compact local slipper snail *Crepidula*.

As he began to study the development of *Crepidula* from its earliest egg cell stage, Conklin observed the morphological details of cell division and nuclear as well as cytoplasmic activity. Conklin found the earliest stages of development fascinating. He turned closer and closer attention to them, thus embarking on what became known as cell lineage work, in which researchers meticulously traced the detailed changes in each cell as it underwent division. They thereby charted the lineage of each cell as it became increasingly differentiated towards its eventual role in the adult organism.

One day, Conklin reported, Columbia University cytologist Edmund Beecher Wilson walked across the street to talk with him. Wilson had also studied with Brooks at Johns Hopkins but had completed his Ph.D. roughly a decade earlier. He had already achieved a reputation with his careful cytological and embryological work. Conklin was thrilled that Wilson was visiting and he was even more excited when he learned why. Wilson had been working on his own cell lineage studies, using the marine worm *Nereis* instead of Conklin's *Crepidula*. Perhaps they could make a close comparison of the two organisms and their changing cleavage patterns. Such a

E. G. Conklin in 1920s. Photograph by Julian Scott, MBL Archives.

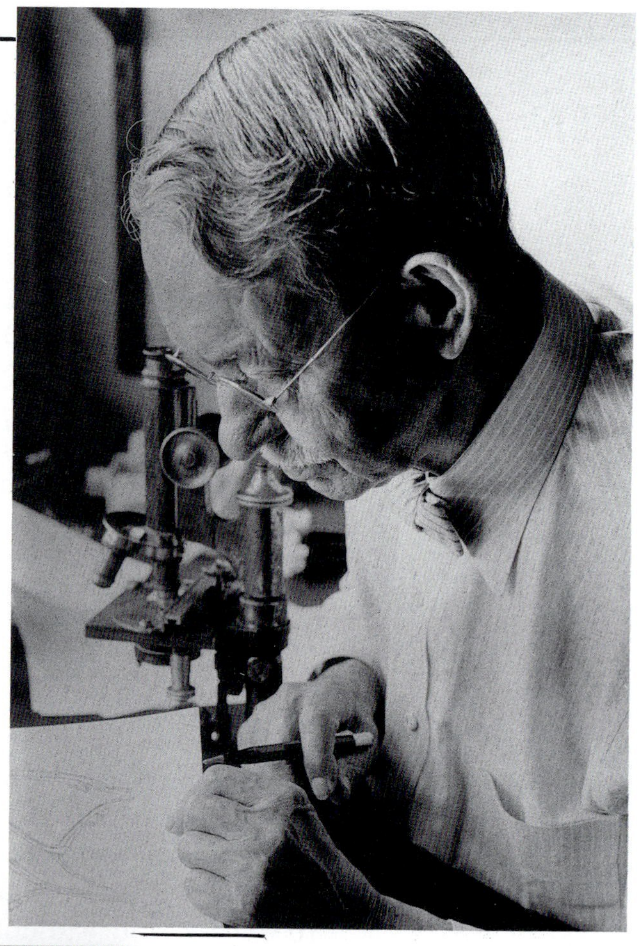

Edmund Beecher Wilson, cytologist and MBL trustee. Photograph by Alfred F. Huettner, MBL Archives.

comparison might reveal useful keys to evolutionary relationships among the simple marine organisms, which might in turn help to illuminate those pressing traditional morphological questions about the origins of the vertebrates and of humans. Conklin and Wilson did compare results and found striking and informative parallels. Many of the cell divisions proceeded in nearly identical ways, with deviations occurring as needed to produce the structural differences in the two adult organisms.

Wilson also reported that Whitman wished to talk with Conklin about his work. Whitman had been one of the first to suggest that looking at the early cell stages might be useful to answer important evolutionary questions, and he was interested in what he had heard of Conklin's work. Conklin went when invited, of course, and explained what he was doing. Whitman asked what he intended to do with the work, and Conklin responded that he did not know. In fact, he rather despaired of finding someone to publish his dissertation, especially since Brooks had said that publication of Thomas Hunt Morgan's dissertation the year before had used

Thomas Hunt Morgan, 1923, the first Nobel laureate with MBL connections. Photograph by Alfred F. Huettner, MBL Archives.

up all the Hopkins money available for such purposes. Whitman offered to print it in the *Journal of Morphology*, which he edited, thereby solving Conklin's problem.

But Whitman insisted on color plates with great detail to show every aspect of cellular development, which entailed considerable careful preparation and additional expense. It took Conklin six more years before his work finally came out in print in 1897. All the colored plates and the sheer length raised the cost of printing to the remarkable sum of $2,000. At one point, when Conklin feared that this meant the end of the project, he expressed his concern to Whitman. Whitman saw no problem, characteristically responding "After all, what is money for?"

Whitman's support, and the community of sympathetic workers that he found at the MBL demonstrated the vital way that the MBL influenced science. Other researchers soon joined in the cell lineage work, and Conklin moved across the street to become a loyal MBL investigator and later trustee until his death in 1952. Whitman taught students and colleagues to look closely, to see details, and to draw carefully. Conklin's own teacher, Brooks, ridiculed Conklin's masterful job, suggesting that he did not see the point of such mere "cell counting." Yet Conklin did not just count cells, and Brooks came to accept Conklin's own insistence that he was a "friend of the egg"—the whole egg, both cytoplasm and nucleus together. Conklin's view had its influence on later embryology classes, which had become increasingly specialized and reductionistic by the 1930s. Students recall Conklin's decades of visiting lectures and his dynamic emphasis on the egg and on the whole cell and cell interactions.

The Tradition of Research

Costello later adopted Conklin's traditional emphasis when he undertook work on *Nereis*, the same worm that Wilson has studied. Wilson, who suffered from crippling arthritis, died in 1939, but his wife continued to spend her summers in Woods Hole. When the young Costello met her and explained that he was studying the embryonic development of *Nereis*, she responded happily, "Oh, my husband loved that egg." But, she reported, the problem was that with *Nereis* one has to collect specimens and then start observing at night. That meant that even the year they were married, he was spending his nights collecting and watching in the lab. "That," she recalled, "was very distressing." Years later, when Costello met her again at a party, she recognized him as the young *Nereis* fellow and asked what he was doing. When told that Costello was still examining the nature of cleavages in the *Nereis* eggs, Mrs. Wilson responded that "E. B. would have *liked* that." A classic paper of Wilson's on germinal localization in the molluscan tooth shell *Dentalium* appeared in the first volume of the *Journal of Experimental Zoology*; Costello's paper on *Nereis* is in the one hundredth volume. Clearly, Costello felt very proud to belong to such a tradition of cell lineage studies begun by Wilson and Conklin so long before at the MBL.

In fact, he insisted to me when I was a fresh graduate student in the history of biology that I should study *Nereis* myself. But he did not want me ordering eggs or worms from the supply room. Rather, it was essential to take Costello's own net and his light and obtain special permission to go to the MBL dock at night. There in early summer, shining the artificial light into the water mimics the full moon and stimulates the beautiful and complicated mating maneuvers. The male and female worms begin to dance toward the light. They trace out intricate spiral patterns in the water, becoming more and more frenetic with time. Eventually the moment comes when the females release their eggs and the males release their sperm into the water, which suddenly becomes very cloudy. The embryologist can either grab the worms just before that point or collect some of this water and run for the lab to start watching cell division. The realization that simple observations are extremely difficult to make, and that it is very tedious to try to reproduce the meticulous cell lineage work that was done with simple equipment a century ago is instructive. The sense of following the same tradition of research as Wilson, Conklin, Costello, and others is also inspiring. Some recent returnees to cell lineage work have similarly felt the pleasure of participating in such traditions.

Geneticist Tracy Sonneborn epitomized the feeling and the excitement at the MBL when he wrote to his aunt and uncle. He wanted to thank them for their generous twenty-first birthday present and to tell them what he had done with the money. As a student at Johns Hopkins under Herbert Spencer Jennings, Sonneborn had enrolled in a summer course at the MBL. He wrote from Woods Hole in 1926:

> If you could only know the childish delight and wonder and amazement which I have been experiencing the last few nights, you would be certain [that I will never grow up]. I just wrote to mother about it, but I'm going to write it to you because I want to be sure you hear.
>
> I've been staying up over my microscope the last few nights *watching a new living being formed!* Can you believe that? Well, it's absolutely so. I've collected certain marine animals whose mating reactions are governed by the moon; at full moon and for the first quarter thereafter they spawn. It is a simple matter to collect them at night with a flashlight; they are strongly attracted to the light, just like moths and swim right up to you so that they can be scooped out and brought to the lab. When I am ready, I allow them to mate. Right away, I take a drop of water containing hundreds of eggs and thousands of sperm and put it under the microscope. Then I can see the whole drama in all its details.

> The sperm swim up to the egg, penetrate it and fuse with it. This act of fertilization completely transforms the egg so that it's [sic] whole substance undergoes a reorganization. It soon begins to divide: the single egg-cell transforms itself; right before your eyes, gradually but surely, into *two* cells. More rapidly, each of these divides in half and there are *four* cells. Now things happen apace. In a little while there are hundreds of cells developed from the single cell; and wonder of wonders, the organs begin to take form, whip-like locomotor appendages appear. Then you think you're seeing illusions from too much staring and gaping: it can't be true. The things are actually beginning to swim around! But you become convinced that such is actually the case. Can you realize the thrill of seeing the whole process of the formation of a new creature? It simply makes you gasp. Do you wonder, then, that I still feel the amazement and wonder of a child when I see such things?
>
> Aren't you glad that your gift is helping me to buy the instruments that make it possible for me to see such things, and to retain the immortal astonishment of a child at *how* things work, and to keep me wondering why? Perhaps, but this is almost too much to hope for—I may even help to find out the *why* and *how* of some of these wonderful things, with the aid of those instruments.

The only flaw in carrying out these studies today is that the current emphasis on the giant squid axon has introduced some squid into the Eel Pond. These animals like to eat the unsuspecting *Nereis* in the midst of the worms' sexual activities. Ethicists somewhere probably worry about whether the observer is morally guilty for provoking the attack.

Leaders

In those early decades of collecting on the dock, there were a few leaders who clearly stood out. Conklin, Wilson, Morgan, Loeb, Montgomery, Harvard's George Howard Parker and Yale's Ross Granville Harrison joined Whitman and Lillie as top men. To one admirer, it seemed that the senior scientists in the beginning were "real luminaries in science" and made a lasting impression. Later there seemed to be many more scientists who stood out in different ways rather than as leaders overall, but that change was largely due to changes in science and the structure of science rather than to anything about the MBL itself. With time it became much more difficult for individuals to master all the concepts and problems of the day the way those leaders seemed to in the early part of the century. Researchers saw with some regret the moves from general biology, concerned with concepts pervading the field, to what one observer called the more specialized ATP Age and later the DNA Age. In addition, the pressures to "publish or perish" and the greater rewards for publishing ahead of the competition have made many scientists think twice before openly discussing their ideas and innovations, thus cutting off free discourse. Science changes, and so do the individual scientists' roles. Today there are some administrative leaders, some brilliant fund-raisers, and some world-class scientists. Very few people have the time or energy to excel in all areas.

In those earliest years, Whitman, Lillie, and the others could represent MBL leadership both scientifically and administratively. These were all exceptionally strong men, with a conviction as to what a biology program should be like and about how to effect such a program at the MBL. Those were simpler times, when they could feel confident about what they were doing and could urge inclusiveness rather than always being forced to make exclusive decisions. These men towered far above the others in leading the MBL and American biology. They took canoeing and hiking trips together, attended meetings together with long train rides to talk things through, and they edited the *Journal of Experimental Zoology* together. These men were close friends and made up the sort of unique group that rarely comes together. They were also recognized as world leaders in biology generally. Their presence at the MBL helped show the international scientific world that this was a leading place for both research and teaching excellence.

MBL *Personalities*

There were others who stood out through force of personality as well as scientific excellence. Columbia's Thomas Hunt Morgan was highly regarded, for example, not only for his work on regeneration, heredity, sex determination, and later genetics, but also as a popular member of the MBL crowd. He was known, at least to some of the Woods Hole community, for his clean white shirts. The story, greatly exaggerated according to his daughter Isabel but told by several local sources, was that his mother insisted that her son's shirts should be a priority for laundering. He must have clean white shirts, rather than the more typical detachable collars, to wear each day to the lab. This vehement demand impressed the launderers in town.

Even such prominent people could be the subject of friendly jokes. Paul Reznikoff recalled a party at Columbia attended by a number of MBL people. Morgan, who was later to win a Nobel Prize, appeared at the door and asked what people were drinking. Cocktails, was the reply. Morgan asked for one, though it seemed to the partyers that he was not used to such drinks. As he began to feel the effects, he asked Reznikoff, "Young man, why don't you change your name?" "Well, Dr. Morgan, I really don't see why I should," Reznikoff answered. Morgan rejoined with alcoholic wisdom, "Oh, I see, I see; that's your name and you want to keep it." "Well," Reznikoff responded, "that's probably a good reason. Now you tell me, Dr. Morgan,

Robert Chambers, cell biologist. MBL Archives.

why don't you change *your* name?" After all, Morgan was named after a "bloody pirate." The students might have pointed out Morgan's relationship to a Confederate general as well. It is always reassuring to "have something" on the great leaders.

Columbia's cell biologist Robert Chambers provided even more fond amusement to the MBL community. At the end of the season one year, he and his wife were rushing to make the train on time, as the family often ran late. The laboratory community had pitched in to help get all the Chambers's belongings to the station, because very few people had cars in the mid-1920s, and those who did often used them to help others. When they reached the station, the Chambers couple suddenly realized that they had left their son Bobby home in the bathtub. Mrs. Chambers kept the engineer busy talking, while the young helpers hurried back to wrap the baby in newspapers and bring him to the train.

Robert Chambers was evidently the prototypical absent-minded professor. A favorite story concerned a day back in New York, but it could have happened in Woods Hole. Chambers had taken all the family's umbrellas to be repaired and had forgotten to retrieve them. When he went off to

Columbia in the morning, his wife insisted that he must stop and pick up all six on his way home. Chambers rode the street car to work, then absent-mindedly picked up the umbrella next to him as he prepared to disembark. The lady next to him said, "Hey, mister, that's my umbrella." With sufficient apology and a charming smile, he persuaded her that he had simply made a mistake. That evening, he entered the street car with his six repaired umbrellas all wrapped up to take home. There was the same lady. She said, "Oh, say, you had a good day today, didn't you?"

Cooperative Administration in a Community

The MBL has always attracted a variety of people, able scientists with various personality quirks that make the community more interesting. From the beginning Whitman stressed that the MBL was a community, consisting of specialty cells organized into a functional, cooperative whole organism. As second director, Lillie chose to perpetuate that view. Whitman imposed his vision of the MBL by running the place in a dictatorial manner and making as many of the decisions as possible himself. Lillie held far more meetings but brilliantly used committees to work things out. Lillie reportedly had little patience with interminable arguments, where everyone saw the positions early on and realized the points of disagreement. Instead of listening further and working to effect some compromise, he appointed a committee, naming the most vocal people on the extremes of a question and giving them the task of bringing forth a proposal.

The system generally worked and produced the sense of cooperative administration through committee work. However, perhaps it worked too well in the early decades, because it made people want to stay and be part of the MBL. The same people loved Woods Hole, bought summer places there, and returned summer after summer. With so few housing openings it became too difficult for many, especially married, new researchers to join the group. This created a shortage of new leaders by the 1930s, and it took new housing and new resources to attract new blood. Perhaps the same sort of closure could happen again, unless more housing can be provided to keep young people coming back and bringing their own students, who will in turn keep coming back. . . . The MBL is trying to meet this critical need.

Nonetheless, the sense of a community of research has persisted and, indeed, represents the spirit of the MBL. New people have always managed to squeeze in among the "regulars." The decision in 1940 to rotate course instructors more frequently than the nine years then typical, reducing the post to no more than five years, helped to bring in enthusiastic and energetic recruits. The wonderfully appropriate sculpture of a group of three scientists engaged in eager conversation, created by Elaine Pear

Cohen and entitled "Woods Hole: The Scientists," depicts the atmosphere. And the story from 1929 about Maurice Rayon, the botanist who worked on silkworms, illustrates the importance of community. Someone heard that an unpleasant person was coming to occupy the one empty research table, so the botanists invented Dr. Rayon and kept his correspondence and research going for a while until the crisis had passed. The table was therefore "occupied" and safe from the undesirable visitor. No one knew just who had done what or who was in the know.

Today the institution has become too successful and too large to rely on such a cooperative, community approach any longer. Those who call for a return to the old times probably do not understand the complexities of funding and publicity central to any modern research institution. Those who long for those old times have many sympathizers, but perhaps little foundation in reality. The current system of committees and helpful administrators at all levels has made many things run more smoothly.

Jobs and Fellowships

One thing that has not changed appreciably is the diversity of people attracted to the MBL, some bringing with them little more than a desire to be here. One young student, Arnold Clark (later a geneticist at the University of Delaware and a long-time MBL regular), was so eager to learn biology that he arrived with no job, only ten dollars, and no place to stay. He went to the head janitor, T. E. Tawell, to ask for a job. Tawell replied that things were very tight just then and asked whether Clark could wait for a while, maybe two days. In two days, Clark returned and was given a job sweeping the floors of Lillie at five in the mornings. He would arrive at that time, have a glass of grapefruit juice with Mr. Tawell, then get straight to work, finishing in time for his morning class and laboratory work. One morning he had to return to the Lillie Building in midmorning. To his dismay, another fellow was sweeping the very floors that he had swept only a few hours earlier. He dejectedly asked Mr. Tawell what he had done wrong. "Nothing," was the answer as Tawell explained, "Well, Arnold, you're not the only one around here who needs a job."

Others found jobs in the collecting department. When Horace Stunkard arrived from the midwest, where he had been told that he must go to a marine station for a season in order to receive his degree, he had no job either. He applied to George Gray in the supply department and obtained a position helping with collecting, despite his landlubber's ignorance of marine organisms. He learned fast; he had to.

Fortunately, various granting agencies have arisen in the course of the twentieth century, so that investigators now arrive with National Science Foundation (NSF), National Institutes of Health (NIH), or even Guggenheim

And still they sweep. Janitor sweeping out Lillie auditorium, 1952. MBL Archives.

Polly (foreground) and Ruth Crowell, longtime MBL employees. Photograph by Alfred F. Huettner. MBL Archives.

Fellowships under their belts. Students may apply for a number of special scholarships to attend courses or may receive fellowships to carry out research. MBL advocates have recognized the importance of such funds for scientific work, as when Columbia physiologist Harry Grundfest appealed to the community to write to congressmen to urge passage of the NSF bill. And generous donors have set up important privately funded fellowship programs.

Just one of several such programs that reflect the MBL spirit so well is the Frederick B. Bang Fellowship. Bang, a physician, was Professor of Pathobiology at Johns Hopkins, studying infectious diseases. On his way back to Baltimore from doing research at Mt. Desert Island in 1950, he stopped in at Woods Hole to visit a colleague, who insisted that he and his family stay for a few days. They did and were hooked from then on. When Bang visited the Marine Resources Building, he was astonished that they simply threw away the dying animals, which were of no interest to the researchers who needed healthy living specimens. With his interest in pathology and immunology, he was amazed at the waste. He saw a potential gold mine of research material in these discarded animals. The next year, the family returned to Woods Hole, bought a house, and continued to return thereafter. Bang and his wife, Betsy, loved their work at the MBL. Thus, when Bang died in 1981, the family sought to make it possible for other researchers to work at the MBL on the immunology, pathology, and infectious diseases of marine invertebrates. Their story is typical of the sorts of cross-fertilization of disciplines that occur here, and the ways in which the effect carries on.

Another such fellowship was the Lucretia Crocker Grant, named after an influential Boston educational reformer, and designed to help teachers learn natural history. This grant program provided an opportunity for many women at the MBL, as many of them were school teachers. With fewer high school teachers coming to the MBL, the Crocker fellowship recipients have changed a bit over time.

Minorities at the MBL

Only recently have larger numbers of women begun to enter the domains of investigation and instruction in their own rights rather than as students or as adjuncts to husbands and advisors. At first not all of the women were well received. Some were the brunt of unfriendly practical jokes. For example, a group of men saw one of the women, a rather hefty person, on the floating collecting dock with her heavy and voluminous skirts on, bending over to haul up a sample. They casually stepped together onto the dock, which thus sank several inches and soaked the poor woman, whom they did not really like. She probably did not like them much either. Others,

Sister Florence Marie Scott, 1964. She was a corporation member and trustee from 1902 to 1965. Loaned by Jane Fessenden.

such as Cornelia Clapp, have been more fortunate and more integrated into the community since the very first years.

Examples of women researchers from the middle of this century include the popular and respected Sister Florence and Sister Elizabeth. Both spunky and dedicated researchers, they have inspired many fond reflections. Sister Florence, who returned to the MBL for over thirty years, was so popular that she once said that she almost dreaded walking through town. She knew so many people and always stopped to chat so that it took a very long time to get anywhere. When George Scott insisted on calling her

E. E. Just in Old Main, 1923. Photograph by Alfred F. Huettner, MBL Archives.

"Sister Scott," since that was her last name, she retaliated by calling him "Great Scott." She had a fine sense of humor and told wonderful and not always proper stories. She and Sister Elizabeth helped to break down any prejudices that nuns could not do science.

Sister Elizabeth was also admired. Early one morning she went to the MBL floating dock behind the supply building to check on her organisms. When she stepped on a plank, it somehow rolled and she fell into the water. Nobody was around, so she calmly grabbed onto a nearby piling and hung on until someone came to rescue her. The collecting crew solicitously insisted that she be taken home in a truck, though she did not want to get it all wet. When she arrived at the house where she was staying, Mrs. Smith kindly helped her into a hot bath and took care of her. Sister Elizabeth explained that she appreciated all the attention, but said "You know what I *really* would like to have had would have been a drink of whiskey!"

Black students and researchers have also found a place at the MBL, though not perhaps in the numbers or quite as early as they might have. Kenneth Manning's study of Howard University embryologist Ernest Everett Just and early twentieth-century times at the MBL indicates how uncomfortable Woods Hole could be for blacks in the teens and twenties. It was, of course, a reflection of racism in the society as a whole, carried over to the MBL community as well. It is clear that many scientists held the highest respect for Just's biological work on fertilization and the cell surface. As Paul Reznikoff recalled, Just was "one of the most remarkable men" he had met, a meticulous researcher who was always ready to give up his own valuable time to help anyone with experimental work. While Reznikoff's group was getting about 60 to 70 percent successful cleavage, Just would get 98 to 99 percent. They asked him how he did it. In his typical way, he showed them. He kept his starfish and sea urchins in a covered bucket, even during the very short time it took to move them into the lab. He thus avoided the accelerating and confounding effects of the sun.

Foreigners have always been officially welcome, though the relative proportions of visitors from different countries have changed significantly with time. The wars and depression brought refugees to America, among them many excellent scientists, but the MBL did not have sufficient resources to take them in in substantial numbers. Refugees often arrived with nothing to support their research or even to live on, but helpful scientists did sometimes adopt some of them and gave them space in their labs.

Japanese students have played a prominent role at the MBL at various times, beginning in the earliest years when Whitman's student Shosaburo Watase attended. Whitman regarded Watase as probably the leading cytologist in America and promoted him at the University of Chicago and the MBL. Spending his summers in Woods Hole, Watase evidently felt himself well treated and comfortable, despite his imperfect English. In Chicago, he met with less hospitable treatment as the administration criticized his teaching and failed to award him promotions. This leading cytologist eventually left the United States and the MBL to return to Japan.

In the early years Whitman also hired Japanese artists for the summer staff, to help with drawing and final coloring of plates. Eventually they found American artists and photographers as well. Whitman's connection with Japan dated to the years he had spent teaching biology at the Imperial University of Tokyo. He had had only four graduate students and fairly limited facilities to work with, but all four became published, successful professional biologists. Whitman gained tremendous respect for the Japanese people but not for the University of Tokyo bureaucracy. He left his job in Japan partly because they told him that his students could not publish their own work under their own names. The professor's name must be given as author. Whitman rebelled against what he saw as a gross stupidity and helped to have the papers published in other, more tolerant journals elsewhere. In Japan, he nonetheless acquired a love for the people that set the atmosphere of acceptance at the MBL.

A framed notice on the wall of the library's card catalog room also tells of the sympathy between Japanese scientists and the MBL. The handwritten sign reads:

> This is a marine biological station with her history of over sixty years.
>
> If you are from the Eastern Coast, some of you might know of Woods Hole or Mt. Desert or Tortugas.

If you are from the West Coast, you may know Pacific Grove or Puget Sound Biological Station.

This is a place like one of those.

Take care of this place and protect the possibility for the continuation of our peaceful research.

You can destroy weapons and the war instruments.

But save the civil equipments for Japanese students.

When you are through with your job here notify to the University and let us come back to our scientific home.

—*The Last One to Go*

This had been addressed to the submarine squadron occupying Mikasi Laboratory. When Costello saw the poster, which had found its way to the Woods Hole Oceanographic Institution, he thought the brushed handwriting looked familiar and had a friend look up an old Japanese friend of his from the MBL, Katsuma Dan, to learn if he had made it through the war. He had. And Costello was right, the poster was Dan's.

Katsuma Dan had come to the United States to work on a Ph.D. with cell biologist L. V. Heilbrunn, even though Dan knew little English. Dan and his wife, Jean Clark Dan, were very popular, so that the MBL community sympathized as the Dan family in Japan suffered various political problems.

This is a marine biological
station with her history of
over sixty years.
If you are from the Eastern Coast,
some of you might know Woods Hole or
Mt Desert or Tortugas.
If you are from the West Coast,
you may know Pacific Grove or
Puget Sound Biological Station.
This place is a place like one of these,
Take care of this place and protect
the possibility for the continuation
of our peaceful research.

You can destroy
~~the~~ weapons and
the war instruments
But save the civil equipments
for Japanese students
When you are through
with your job here
notify to the University and
let us come back to our
scientific home

The last one to go

Sears Crowell and Emperor Hirohito examine some hydroids and nudibranchs, 1975. MBL Archives.

Japan had been a strictly closed country, and Dan told Costello that his father and uncles were brought up to believe that they would be beheaded if they went into a foreign country. Japan was opened when his father was about sixteen, however, so he was sent to study engineering in the United States, where he and his brothers attended Harvard and MIT. They then returned to introduce American engineering and mining methods to Japan. When Katsuma Dan arrived in turn years later, he and Costello shared a lab at the University of Pennsylvania. They drove together to California one year, in a car that boiled over at every hill, and they then went on to the MBL together. At the University of Tokyo Dan later taught cell biologist Shinya Inoué, now a leading year-round MBL researcher.

After the Dan era, the emperor came to the MBL. In 1975 Emperor Hirohito proposed to visit the United States. Where did he wish to go? To the Marine Biological Laboratory in Woods Hole, undoubtedly leaving a host of diplomats scratching their heads and scurrying for their Massachusetts maps. A marine biologist himself, he wanted to see the famous laboratory. But the MBL is not accustomed to fancy dress and formal company. They fixed up the men's room, which had languished away for years and had acquired a layer of grime and intellectual graffiti. They polished the floors and covered them with red carpets. They prepared tea for the honored visitor and presented an impressive array of fancy local cookies. The emperor came, took part in formal receptions of WHOI and the MBL, and graciously signed his name in a guest book and in one of his own publica-

Emperor Hirohito autographing a copy of his book, Some Hydrozoans of the Bonin Islands, *in 1975. Left to right: Paul Fye, James Ebert, and secret servicemen.* MBL Archives.

tions (now in the Rare Books Room). When the time came for the scheduled tea break, the emperor instead rose and said "Let's get on with the science." The group then looked through the microscopes at the hydroid specimens that had been prepared by Indiana University biologist Sears Crowell and set up in the library. Security remained tight around Woods Hole and especially in the MBL library that Saturday, and people brought out and fit themselves into suits they probably had not used at the informal MBL for many years.

W. J. V. Osterhout reports a rather different sort of special visit years earlier when President William Rainey Harper of the University of Chicago came to visit around 1900. Chicago was a major research university and an important supporter of the MBL. Harper was quite a dignitary by MBL standards of the time. Captain Veeder, who took charge of boating expeditions, arranged a clambake at Naushon Island for the distinguished visitor. One of the guests was introduced as a general in the Grand Army of the Republic. He pulled out his flask in that late nineteenth-century time when few proper people imbibed (at least in public) and proceeded to become quite genial. He even became sufficiently playful to throw a slice of watermelon, which promptly landed on Harper's stiff formal white shirt. The next day Jacques Loeb was to take Harper to the train. Since they arrived early, Harper decided to get a haircut. He could not understand why Loeb so strongly resisted his having his hair cut. Finally Loeb broke down and escorted Harper to the barber, who was that very general, in no shape to cut anyone's hair.

In another incident a photographer concluded that Loeb was trying to poison him and vowed never to return to such a crazy place as the MBL. This man was helping Loeb with some difficult photographic work. He insisted that he was ill and could only work if he had some whiskey to help him. Loeb recalled that he had seen a bottle in Whitman's lab with a well-known visiting scientist's card of thanks attached. Because Whitman did not drink whiskey himself, Loeb assumed that he would be willing to oblige the photographer. Whitman was happy to do so. It turned out, however, that the visiting scientist had never really left any whiskey. Instead Lillie and Parker had filled some old bottles with seawater and left them on Conklin's desk with a leftover card of the scientist's. Conklin had recognized the joke and had responded that there must have been some error and that the bottles were surely intended for Whitman. Thus they had been carted off to Whitman's lab, where Loeb found them. Salt water it was, and Loeb had to find other help to complete his photographs.

E. G. Conklin and son in a moment of relaxation together. MBL Archives.

E. B. Wilson with sons of the photographer. Photograph by Alfred F. Huettner, MBL Archives.

MBL Staff and Children

In addition to the scientists, the administrators, and the librarians, the MBL has a host of other people who keep things running year-round. These people are truly indispensable, even though they are often not accorded the attention they deserve. From the watchmen to the janitors, to the collecting, dining hall, electrical, buildings and grounds, chemical, and mail room staffs, and all in-between, this considerable and influential group includes individuals who are fascinating in their own right. One finds a buildings supervisor who was born and has always lived in the same house in Woods Hole and whose family has worked for the MBL since the beginning. Or a watchman who has traveled around the world and has read more widely than most academics. Or the administrative assistant whose grandmother owned houses next to the Eel Pond and who herself worked cleaning rooms at the lab years ago. Or the janitor who, as a native Latin American, has a wealth of information and more than a few ideas that he would like to teach the U.S. government officials about Central American life, history, and politics.

One other group of MBL people to whom visitors should listen if they want to get at that much-touted MBL spirit is the children. MBL children are

Lilian and Isobel Morgan with chicken. Photograph by Alfred F. Huettner, MBL Archives.

T. H. Morgan with daughters Lilian and Isobel, 1918. Photograph by Alfred F. Huettner, MBL Archives.

unique. Recently two young boys, probably about ten years old, were talking at the MBL beach. One reported that his family was going to Australia in the fall for his dad's sabbatical, and that it was going to cost them each over $1,000 for airfare. But that was okay, he said, because they were going to take it off their grant. The other boy asked why they were going. "Well," came the ten-year-old's reply, "the animals are different colors there and they develop differently than the ones here." The other boy reported that his family was just going to North Carolina for their leave the next year. "That's not so bad," the first one replied, "we went there once for my mom's sabbatical and it was neat too." Children do not talk like that in the less stimulating "real" world.

Neither do children out there choose to spend their summers in school, let alone to study something serious like science. But many MBL families find themselves drawn back to Woods Hole even in years when they meant to stay home or go elsewhere. The children do not want to miss their friends or a session of the Science School, held in the old schoolhouse building. There they take classes of all sorts, with a liberal dose of seashore field trips included. Ten-year-old Science School students know more about marine organisms than many professional biologists in other parts of the country. Then as the children get older, they often work at the MBL, maybe in the supply department, as Whitman's and Conklin's sons and so many since have done, or in the Mess, or cleaning dorm rooms, or more recently, in the library or photocopy room. Perhaps it is the enthusiastic response of the children that captures and perpetuates the MBL spirit best.

Unknown child on the beach. Photo by Alfred F. Huettner, MBL Archives.

FiS kids: Rebecca Jackson, Tracy Goldsmith, Eric Hallstien, John Gagnan. Photograph by George Liles. MBL Archives.

The new "FiS Kids" program to promote Futures in Science for high school students and their teachers promises to involve more of the local youth in science at an early age. During the first year, the four selected students pursued sophisticated projects. Each got negative as well as some positive results. Each expressed with enthusiasm the value of the experience. These students learned early what science is about: it is not always so neat and tidy as it seems from most preprogrammed elementary laboratory exercises, but it is a lot more fun and exciting. With the cooperation and support of the MBL Associates, the Commonwealth of Massachusetts, the Monsanto Corporation, and the Falmouth school district, this program exhibits a remarkably successful cooperation of public, private, and institutional interests. "Get the younger generation involved in science, and keep science young" is the motto here.

NOTES

Donald Costello, discussed his first arrival at the MBL in his interview recorded in the MBL Archives.

Conklin, on his first years at the MBL, in "Early Days at Woods Hole," *American Scientist* (1968) 56: 112–120, as well as in other writings and even a phonograph record made during an interview at Princeton toward the end of his life in the 1950s.

Conklin loved talking about what science was like in the early part of his career. In a more formal account, Conklin discussed William Keith Brooks in his Memoir for the *National Academy of Sciences, Biographical Memoirs* (1910) 7: 25–88. He always considered himself a "friend of the egg—the whole egg," as many students have recalled. See also J. Bonner, with notes by Whitfield J. Bell, Jr., " 'What is Money For?': An interview with Edwin Grant Conklin, 1952," *Proceedings of the American Philosophical Society* (1984) 128: 79–84. Conklin's, as well as Wilson's, Morgan's, Whitman's, and other bound collections of reprints, including their classic early papers, are housed in the MBL Rare Books Room.

There were, in fact, other "top men" besides those named here, but these men returned year after year, served as trustees for long periods of time, worked as journal editors, and generally continued to play the most active roles in all facets of MBL life for a long period of time. More information about each of the men is available in the archival collections at each home university or at such central collections as the American Philosophical Society.

Robert Kahler's Historical Collection interview, recalls Morgan's white shirts, though Isabel Morgan Mountain questions whether the story was overstated.

Tracy Sonneborn, letter of August 2, 1926, to his Aunt Bella and Uncle Jake, printed in *Collecting Net* (July 1987) 5: 14.

Paul Reznikoff's Historical Collection interview discusses the cocktail party and Chambers's stories. Many people fondly recall Chambers's superb idiosyncrasies, often chuckling while relating stories so many years later.

University of Chicago records as well as the Lillie Collection at the MBL demonstrate Lillie's management style. Sears Crowell provided an account of Maurice Rayon, the phantom botanist, in a story told by Hannah Croasdale, MBL Archives.

Arnold Clark told the janitorial incident in an interview while in the MBL Archives one day, while Horace Stunkard recorded his arrival at the MBL in his interview in the Historical Collection.

Betsy Bang, personal communication in 1987 and 1988, provided information about the Bang family's arrival in Wood's Hole and about the Frederick B. Bang Fellowship Fund.

Sister Florence and Sister Elizabeth are remembered in George Scott's interview, Historical Collection, as well as by Donald Costello and a number of others.

The fullest discussion of Ernest Everett Just's stay at the MBL appears in Kenneth R. Manning's extremely well-written biography, *Black Apollo of Science* (New York: Oxford, 1983). Whitman's experiences in Japan are discussed in biographies of Whitman as well as in his unpublished account "Zoology in the University of Tokyo," all available in the MBL Archives.

In his interview deposited in the MBL Archives, Donald Costello recalled his times with Katsuma Dan, and a number of documents recall Dan's role at the MBL.

During Emperor Hirohito's visit, James Ebert was director of the MBL. Records in the MBL Archives and Ebert's own recollections show that, even with limited resources, he managed to pull off the event gracefully and without problem—no easy task for an informal place without real precedent for state visits.

W. J. V. Osterhout discusses University of Chicago president Harper's earlier visit and the whiskey bottle, in his Loeb stories sent to the MBL Archives in 1948. Also see Conklin, "M.B.L. Stories," (Notes, Chapter 2: "The Story of the Whiskey Bottles," pp. 128–129.

6

Doing Science

Tarpaulin Cove, lighthouse and bell, 1896. Photograph by Baldwin Coolidge, courtesy of SPNEA, Boston.

Captain Veeder aboard the MBL collecting vessel Cayadetta, *1923.*
Norman W. Edmond Collection, MBL Archives.

LABORATORIES TODAY house all sorts of exotic-looking paraphernalia. In some cases, the equipment even spills out into the halls, with large, heavy-duty, impressive-looking apparatus designed to sterilize or rotate or otherwise manipulate the required materials for the intricate operations that make up science. Fancy high-powered computers have appeared, for example. Or Shinya Inoué's unique six-foot light microscope, which is both large and complex, but which makes it possible to observe details of living organisms instead of the frozen and prepared dead specimens required for the more typical electron microscopy. Elsewhere, bright yellow signs on some doors declare the radioactive goings-on inside. Much of twentieth-century biology has become complicated and expensive, often requiring teamwork. Of course, this does not keep the researchers from personalizing their space with posters, favorite photographs, or even afternoon coffee or tea breaks.

Afternoon tea poured by Mrs. Albert Svent-Györgyi. MBL Archives.

Collaboration

Scientists often find that they need a particular tool or a particular procedure which is not readily available. They cannot always just locate the item in a handy mail order catalog and phone in an order. So the scientist becomes part designer and engineer. And each MBL researcher learns from the others. One cell biologist reports that his work in embryology depended on finding a functional probe of a particular sort to perform delicate operations. Only in talking to a neurobiologist on the MBL beach did he realize how to solve his technical equipment problem. Other examples abound: there is the neurophysiologist working on the horseshoe crab *Limulus* exchanging ideas with the expert studying the workings of the fish retina; they collaborate on analyzing what the brain tells the eye to do. The salt-marsh ecologist learns from the physical chemist about uptake of various chemical nutrients. Or the cell biologist studying cell motility inspires others to look more closely at the effects of osmotic pressure. The light microscopist with his unique equipment makes it possible for the expert on growth of the sperm's extending acrosomal tip in fertilization and

An early laboratory. MBL Archives.

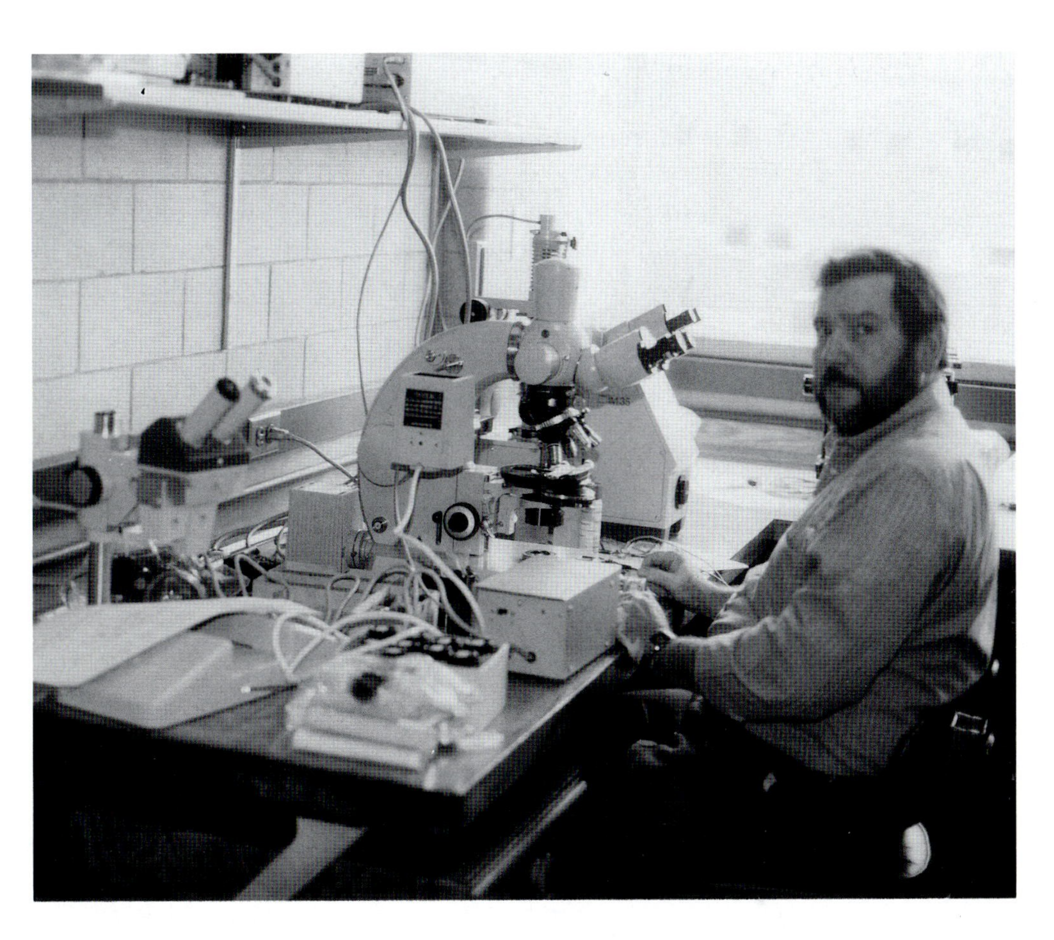

A 1980s laboratory. Photograph by M. Rioux, MBL Archives.

development actually to look and see the tip moving and extending. These research collaborations are possible here because of the unusual mixture of people who might not normally even meet each other, let alone talk and exchange technical ideas.

Indeed, this sort of cooperation makes the MBL what it is, a sort of extended "laboratory without walls," to borrow Albert Szent-Györgyi's term. While scientists elsewhere in the world are sometimes cautioned not to discuss their work in order to avoid being scooped or to avoid priority disputes, the MBL has always thrived on open exchange among a growing group of scientists and students. Back in their academic winter homes, a neurobiologist or embryologist may be able to talk with only one or two other people who care about the same general questions. As Yale cell biologist J. P. Trinkaus put it recently, one tremendously exciting thing about summers at the MBL is that so many of the top people in neurophysiology; cell, molecular, and development biology; and a few other fields are gathered together. The researcher is bound to find someone who understands and will share ideas while walking down the hall, to the beach, or to the library. As neurophysiologist Robert Barlow points out, "If something is not working, 99 times out of 100 someone here at the MBL will have something to help." The absence of teaching responsibilities and of many of the more mundane administrative chores allows researchers intense concentration on those problems of shared interest during the summers. MBL workers seem to collaborate far more often with other MBL scientists than with their longer-term colleagues back at their winter quarters. And they generate an astonishing array of well-attended special symposia and colloquium series all summer long.

Late in his life, Lillie worried that the intense summer research away from the home lab encouraged quick descriptive work and did not allow the careful development of more sustained research programs, particularly demanding experimental programs. It is a problem to pick up and move for the summer, but the "intellectual rejuvenation" and the opportunities for exchange of ideas and techniques, as well as the ready availability of fresh marine material, far outweigh the disadvantages.

Early Days of Marine Biology

The sophistication of the problems and of the equipment today is a far cry from the earliest days in marine biology, when the Frenchmen Henri Milne-Edwards and his friends Victor Audouin and Jean Louis de Quatrefages took to the seashore with only their nets, lots of baggy jacket pockets, and an assortment of collecting bottles. German physiologist Johannes Müller then took his group of students to North Sea holiday spots and added a small boat and larger collecting nets in order to make what he called pelagic sweepings. With such simple equipment, a few men sought to learn about life in the sea, life that they had only recently begun to recognize for the richness and diversity it exhibited.

One of the students who made the pilgrimage to the North Sea with Müller was ardent evolutionist Ernst Haeckel, who sought to find the earliest life forms in the ocean. In turn, he too took his students to the seashore. In his desire to establish a permanent marine station, one of those students, German Darwinian Anton Dohrn, founded the Naples Zoological Station in Italy in 1872. Dohrn did not envision any simple system of nets and collecting bottles. His grand laboratory instead contained a magnificent public aquarium downstairs and wonderfully modern research laboratories upstairs. Visitors could gather there from all around the world, as long as they had a serious research purpose and as long as they or a sponsor subscribed to a table for their use.

Our friend Whitman was the first American at the Naples Station, though as a guest of Dohrn's, since no Americans had subscribed to a research table yet. Then came Emily Nunn, later Whitman's wife. Wilson arrived a few years later, followed by a steady stream of other Americans,

Hats on, hats off: instructors, 1892. Seated: William Morton Wheeler, W. A. Setchell, Charles Otis Whitman, Hermon C. Bumpus, Sho Watase. Standing: Pierre A. Fish, Jacques Loeb, Edwin O. Jordan, Charles L. Bristol, Edwin G. Conklin. MBL Archives.

many of them MBL researchers as well. Whitman, Wilson, and others advised their students to venture to Naples, even after they had their own MBL to satisfy their research needs. For Naples afforded two real advantages, which complemented what the MBL offered: a different set of organisms to work with and an international group of scientists using the absolute foremost of contemporary research techniques. Dohrn made sure that the Naples station had sufficient funds to guarantee the best equipment and materials needed. As a result, in the early years, biologists said, "For techniques, go to Naples." Very soon, the MBL became a mecca on its own for the latest in techniques and ideas.

For techniques, Whitman determined after his trip to Naples that the Americans needed a manual for marine biology. As usual when he perceived a need, he met it. In this case, he wrote the handbook in question, concentrating on microscopic techniques for preparation, preservation, fixing, and staining of specimens. With its emphasis on detailed cytological study of fixed materials, his *Methods of Research in Microscopical Anatomy and Embryology* of 1885 offered the latest word at the time. E. E. Just's methods book or Donald Costello and Catherine Henley's handbook or W. D. Russell-Hunter's introductory texts for study of marine invertebrates filled a similar need for students of embryology and invertebrate biology years later. In fact, a significant number of the standard biological textbooks have been written by scientists affiliated in some way over the years with the MBL.

Science at the MBL looks much different today than it did in those earliest years, and yet there is a continuity to the changes that provides perspective on each shift of emphasis. The return of the same people in different but related roles year after year, and the continuing concern with some of the same problems and organisms has ensured the persistence of MBL traditions. Against that background of stability (at times embryology has dominated; at other times, physiology), the problems, equipment, and other details have become more specialized, more sophisticated, and more expensive. Techniques and equipment, problems, and organisms make up scientific research.

Techniques and Equipment

In the early years, a microscope with the latest lenses and dissecting stand, a collection of glass dishes and slides and plates, the appropriate stains, a microtome to slice up specimens (which could be shared with others), and a few other odds and ends made up the basic equipment kit. By concentrating the sorts of research around several central areas of concern, the laboratory could provide what people needed even with their increasingly complicated demands. Keeping the researchers and students well supplied has always been a priority, and a continual challenge. Much of the official correspondence has concerned keeping up with the growing demands for supplies and equipment.

Something as apparently simple as running seawater in the laboratory is necessary to keep specimens alive, for example. Yet keeping the plumbing system working with 500 gallons of seawater running through the pipes every minute (with its potential contaminants and corrosive salt) has always proven difficult. As one of the men responsible for keeping the system working recently put it, "A valve just won't work with a starfish the size of your hand sitting inside."

Another problem is getting and keeping male and female organisms separated when they all come in together in the same bucket. The researcher wants to control development and to start fertilization when he or she is ready, and not when the animals want to begin. This control makes it possible to observe any chosen stage of development, even the earliest, including the fertilization process itself. For one hundred years, many MBL researchers have concentrated on these early processes, to determine both the patterns and the incredible processes by which two tiny germ cells manage to become one complex, coordinated adult organism.

Further equipment helps to turn the individual scientist's observations into a public record in the form of drawings or photographs. Initially, familiar pens (often crow quills) and India ink made the requisite drawing possible. Others used the simple but effective camera lucida to make their drawings, until photography replaced much of that work in this century. One student reported that he had had to identify and draw by hand clearly one hundred specimens of protozoa for Gary Calkins' protozoology course. The first fifty or so were easy, he said, but the last ten were awful to find, identify, or to represent accurately.

Making observations takes some equipment, whether a basic compound microscope (for only a few individuals have ever used simple microscopes for serious scientific work, even when those would have been just as good) and stains and drawing paper, or maybe an electron microscope and advanced videotaping equipment. Since the 1970s, Carl Zeiss, Inc., has helped to provide the very best microscopic equipment for MBL students and researchers. They loan equipment in exchange for the chance to try out new ideas with the demanding and sophisticated but respectful audience. This program has helped to continue the MBL tradition of making available the very best equipment, which from the very first years has featured Zeiss equipment.

Besides microscopes, there are the chemicals, procured from the chemical room. And special glassware, for a while produced by first-rate glassblower from the University of Pennsylvania, James Graham. In recent decades, researchers have added radioactive isotopes and the necessary accoutrements. There are also autoclaves to sterilize equipment, which the research of the early decades here did not generally require; and centrifuges, bigger and faster and more efficient than those Costello used in his first years here. Complicated specialized equipment appears as well. People sometimes have to improvise to meet their particular needs. NIH and NSF grants finance a high percentage of the increasingly complicated requirements of today's workers.

Of course, sometimes the innovations can prove profound while remaining simple. Syracuse University's Robert Barlow notes that his own neurophysiological work on the brain of *Limulus* suffered from the serious problem that he needed to remove the brain yet keep it alive for at least several days. For obvious reasons, it usually dies under such circumstances. Somehow he had to prevent coagulation of the blood without producing toxic side effects. He could not think of any solution until he happened to meet a colleague in the MBL mailroom and described his frustrations. Well, his friend suggested, why not cool a saline solution, cool the brain, and replace the blood with the saline solution to inhibit clotting. Then do the required measurements and studies while the brain stays alive and cool. The suggestion worked, and the very first time brought measurements over a four-day stretch.

MBL techniques and equipment represent an amalgam of the traditional and modern, sophisticated and makeshift. Walking down the halls of the lab buildings and peeking through the doors, a visitor can see tables, chairs, seawater tanks, glassware, and other basic equipment from an earlier century alongside such things as ultracentrifuges and Inoué's fantastic-looking ultimate (for now) light microscope. The more than 100,000 square feet of space for instruction and research are filled with a full range of equipment and with people to use it in the most creative and innovative ways.

Problems

Just as techniques and procedures have constantly changed, the sorts of dominant problems addressed over the years have also evolved. In the earliest years, most researchers worked on a set of closely related problems. They asked "big" questions such as how organisms develop and what their development tells us about their evolutionary relationships. On the first question, which occupied much of their attention, the MBL group held an almost exclusively epigenetic position. This meant they believed that an organism begins as a virtually unformed, though organized, collection of matter and gradually becomes differentiated into the right kind of organism during the developmental process.

Some MBL embryologists, such as Whitman, were intrigued by the sort of quasi-predeterminism that German biologist August Weismann suggested in his chromosome theory of heredity, with its collection of determinants, ids, and idants. They found equally intriguing Weismann's and Wilhelm Roux's mosaic theory of development, according to which the various pieces of the developing organism act as separate bits of of a mosaic, each predestined by heredity to become a particular body part. Yet they still sought an in-between, essentially epigenetic viewpoint; extreme predeterminism remained a horror to most MBL scientists. Morgan, for example, saw as anathema Weismann's idea that inherited germinal material (or germplasm) might actually determine, in some mechanistic and preprogrammed way, how an organism develops. Such an idea is not true science, he suggested, but just pushes the question of how development occurs back to a "shadowy, ancestral past." Others at the MBL declared that Weismann's speculations seemed as futile as "sorting snowflakes with a hot spoon" and could not possibly explain how development occurs. Still others condemned such predeterministic ideas as a "scientific misdemeanor."

To the MBL researchers, it seemed obvious that the internal organization of the egg itself and its response to environmental stimuli direct development. Inherited nature plays a role in development, of course, but so does nurture—as biologists a bit later would have put it. Problems of development and heredity reigned in the first years at the MBL. DNA changed the picture that nineteenth-century researchers had had of the undifferentiated egg, of course, and made it seem more predetermined. Yet many MBL biologists today would still stress, along with Wilson, Conklin, Morgan, Loeb, and others, that development is every bit as important as predetermined heredity in guiding the differentiation process that turns the germ into an adult.

For the first half-century of MBL researchers the question, "does each organism develop epigenetically or by preformation?" could have been rephrased as: does the egg in some way already possess the organization that the adult then assumes? In particular, how much is cell division of the fertilized egg determinate, or fixed, and how much is indeterminate, or subject to change as conditions vary? Cell biologists today pursue similar questions. The laboratory of Raymond Stevens, for example, explores the biochemistry of microtubules, asking what role they play in cell division and cell processes. Shinya Inoué explores in detail the process of mitosis and cell morphogenesis, looking at each morphological bit of the cell. Richard Whittaker's group examines the way in which the genes control cell differentiation. Others at the MBL have extended cell lineage study to other organisms, such as the nematodes, to discover exactly what each cell does at each stage of the developmental process. Each of these laboratory research programs addresses the basic problems and uses techniques developed by MBL scientists over the course of a century.

Rachel Fink provides an excellent example of blending the old and the new with her work on development of toadfish eggs: she became intrigued by the eggs after reading about the work of Cornelia Clapp a century ago. These eggs adhere to a solid surface, which makes them useful for various sorts of experimental work that tests the relative effects of internal and external changes in directing development. Clapp had looked at the importance of gravity for determining the embryonic axis, for example. Fink decided to study these eggs with current techniques to explore twentieth-century problems. Her work demonstrates the MBL tradition in going beyond the earlier work, to combine tradition and innovation in a progressive, productive way.

Traditional views certainly held that environmental factors can exert a major influence in development, just as internal inherited material does. Researchers such as Jacques Loeb, who emphasized the physiological and physicochemical processes of development, began the continued call for a stronger epigenetic and environmentalist position. To Loeb, it seemed that the female did not even require any male input. The egg did not really need any sperm to get it going along its proper developmental pathway; a simple change of salt concentration in the seawater could do the trick by initiating artificial parthenogenesis. Chemical environmental factors gained top billing with him and his entourage of graduate students and assistants well into this century.

Morgan agreed with this physicochemical orientation, as he embarked on studies of regeneration and sex determination right up until 1910. Then his first work in genetics abruptly confronted him with problems of heredity rather than of physiology and development. Others have carried on the physiological programs, including Lionel Jaffe, who uses a vibrating probe to analyze the role of ionic currents in development and regeneration, for example.

Within the context of discussion about the relative importance of epigenesis and preformation, and about morphology and physiology, came other debates about the relative roles of cytoplasm and nucleus in heredity and development. A number of European researchers had begun by 1900 to accumulate data from observations made with improved equipment (oil immersion lenses and effective microtomes, in particular) and techniques (stains, fixing agents, preservatives). These studies suggested that there might be something important in the nucleus. Indeed, that stainable material in the nucleus—appropriately labeled as the chromatin—might retain its autonomy in the course of cell divisions and might have something to do with heredity. Because the only thing inherited directly and as a whole from one generation to the next is the egg, and because this generation of scientists rejected vitalism (see description below), it seemed that the egg must hold all the material necessary for development. Or at least everything necessary to respond appropriately to external influences that direct development. Because development must be some combination of epigenetic response and inherited action (or else evolution could not work), something about the egg must be inherited and organized in some way. Either the egg cytoplasm or the nucleus was important—or both.

That is what the MBL researchers argued in the 1890s and for the next century: that both heredity and environment play essential roles in directing development. Conklin and Watase reminded those who claimed priority for the nucleus that the mitotic apparatus, necessary for the cell to divide, actually resides in the cytoplasm and directs the cells' cleavage patterns. The nucleus cannot survive, divide, or do anything interesting by itself. Even modern biologists might do well to recall that Wilson's great ctyological work and his classic *The Cell in Development and Inheritance* (1896, 1900, 1925) stressed a balanced view of the nucleus and cytoplasm that was characteristic of the MBL. Geneticists began to draw attention exclusively to the nucleus, so that some forgot about the cytoplasm (though Morgan never did). Others, especially the embryologists, became so distracted by the chase of elusive chemical "organizers" (or substances thought to direct development by organizing the material in the proper way) that they ignored the nucleus and heredity. Yet other prominent MBL scientists such as Conklin continually stressed that they remained friends of the cell—the whole cell, and the whole egg, and the whole developing organism. Recent work in developmental biology, some of the best done at the MBL or by MBL alumni, has returned attention to that lesson: do not forget the cytoplasm in the rush to identify inherited bits of material in the nucleus.

Also it is important to pursue both zoology and botany, the MBL has insisted with varying results. From the beginning, the idea to include both carried the day. After all, the trustees wanted to establish a marine *biological* laboratory. Yet those very first years were really dominated by basic development of animals and by invertebrate studies. Despite the trustees' intentions and Whitman's agreement, it took a while before botany really became established, and it has never gained the status that zoology enjoyed. Indeed, the botanist who had declared that the Johns Hopkins University program in biology contained "plenty of lobster, but hardly enough vegetables to make a decent salad" might well have leveled the same criticism at the MBL. A botany course existed for many years, until the loss of its energetic director brought its demise. The head of the supply department, George Gray, made collections of the organisms attached to the buoys near Woods Hole (especially plants) over several years. As there is no lack of interest among students, such a course may well be revived in the future to help gain the balance that the founders said they wanted. A short course on the cell and molecular biology of plants run by coordinators from the University of Georgia has filled that role and has attracted great attention for the last few years. In addition, the various ecology programs, such as the marine ecology course, may replace more traditional botanical courses. Recent work on algae and cell biology of plants also continues the interest in botanical research.

While botany and zoology have been somewhat unequal partners at the MBL, the attention to morphology and physiology has been much more balanced. In the earliest years, morphological work predominated. This meant a concentration on the structure of organisms rather than on how they work in life. Structure could be studied by fixing and staining the organism to determine exactly how it was made and of what parts. Study of function requires devising some way to see inside the living organism, or inventing clever manipulative or experimental ways to gain information about those internal workings.

When Whitman attracted Jacques Loeb to the MBL, they thereby introduced physiological work at all levels of instruction and investigation into the mainstream of MBL work. Whether he was more determined to include physiology or to enlist Loeb is unclear, but Whitman encouraged both with great enthusiasm. He particularly endorsed Loeb's experimental work on artificial parthenogenesis and on regeneration. Though he did not agree with all of Loeb's conclusions or with the popular newspaper publicity about "controlling life," he was undoubtedly delighted that Loeb's successes attracted others to similar research and to the MBL. To the MBL's benefit, Loeb inaugurated the chemical study of developing organisms as well as physiological work, and he attracted a host of medical researchers and medically related issues to the lab. Today neurophysiological study, pursued by many people from medical schools as well as from biological programs, is one of the specialties of the MBL.

Early on, and probably again today, most MBL researchers are mechanists and materialists: they believe that life structures and processes consist of material and mechanical (including chemical) action. They also generally believe that biologists can explain living structure and function in terms of material, mechanical explanations. This does not, however, imply that they are also all reductionists; they do not all insist that life can be explained in terms of the component parts of the organism. In fact, it may be the interactions of the parts, or some sort of wholism, that explains biological functions. During the years from 1910 through the 1930s, one alternative view of living process was vitalism, which gained attention especially through the writings of former experimental embryologist Hans Driesch. Though most biologists at the MBL were not fully persuaded of the vitalistic insistence that something more than mere material and mechanics exists, issues of mechanism versus vitalism were quite hotly debated during that time.

Today less attention is given directly to such metaphysical and epistemological concerns. Most problems tend to be more specifically expressed than they were earlier: in particular, which pathway makes this cell-cell interaction possible? What does this chemical do in this phase of the neurophysiological action? Or maybe: what is the quantitative ecological balance over a small defined area? The problems are better defined. Impatient with the opportunities that nature provides, biologists have turned increasingly to manipulative experimental work. These sorts of questions have recognizable answers in a way that the broader and more far-reaching problems of the 1890s often did not. They yield more easily to what we generally count as progress. But such questions do not evoke animated discussion by the entire community in quite the same way either. Times change. Now the Friday evening lectures still attract people from the community at large, but some people in the audience admit that they only rarely understand what is going on and what the major question is.

Organisms

Sometimes this focus on getting answers to increasingly specific questions has meant that scientists have lost track of the organisms they are studying. Some of the more intently focused of the "squid visitors" may not be able to identify what their chosen research subject eats or when they mate, for example. Perhaps it does not matter that they do not know. Certainly it keeps the brain less cluttered, with more room for biochemical details or neurophysiological modeling. As one researcher put it, "So what! The squid has the giant axon. That is really all that matters"—because it works so well as a model to address the current pressing problems of neurophysiology.

Similarly, Jacques Loeb was rather disinclined to gather animals himself and far preferred to have someone else get his specimens for him. It did not really matter to him exactly where the animals lived or what their feeding habits at the bottom of some pond might be, as long as they provided useful models to answer the questions at hand. He was not attracted to the idea of wading about in the mud or slipping about on the wet algae of the intertidal zones. Other researchers were more inclined toward the traditional concerns about behavior and life-style, as is characteristic of natural historians such as Just, and they ridiculed Loeb's attitude. A true scientist must know his beast, Just insisted. Today, a range of different types of researchers coexist, exploring different sorts of questions with different sorts of approaches and generating the MBL's vitality with cross-fertilization of ideas.

When a student first arrived in the 1890s, people asked not what problem the research would address, but rather what organism. One had to choose carefully because some choices, like Lillie's selection of the freshwater clam *Unio*, sent the researcher off to less pleasant or less accessible sites to collect specimens. Today's easier transportation makes it possible to roam up to Barnstable Harbor to explore life in the mud flats, or to Sippewissett for the salt marsh. Collectors may tap the colder waters or move on to the warmer currents of the Gulf Stream. And thanks to the efforts of private ventures and then the Army Corps of Engineers, since 1914 the researcher has been able to explore life along the Cape Cod Canal, with its fast-moving currents and access to clear, deep water. Lots of starfish find their ways into collecting buckets from the canal.

Choosing which organisms to study can be one of the crucial decisions in biology. Some creatures simply do not like to perform in a laboratory setting. Others only behave in the relevant way in certain months of the year. In particular, embryological studies have to begin in early summer for many marine organisms. For those organisms that are fertile for only a short time, the researcher has to work frantically, collecting all possible material. After the short mating season he or she can then begin a more leisurely process of preparing and observing the materials collected. Occasionally, this means that the hectic season has passed and the would-be observer has discovered a fatal flaw in the preparations that makes the materials useless. A season might thus be lost for serious research and for publishing any papers, on that problem at least. Until very recently, squid have only been available during the summers, because they come close to the shore then and are far easier to collect and bring back than when they migrate further out to sea in the winter.

The selection of organisms for study has changed over the century, of course. At first, logically enough, people went to the seashore to study exclusively marine organisms. Then Whitman began to haul his pigeons back and forth from Chicago to Woods Hole to carry out behavioral studies on birds as he had earlier studied leeches and freshwater mudpuppies. Morgan turned to the abundant and highly variable fruitfly *Drosophila* for study of sex determination and heredity. The subsequent explosion of interest in genetics encouraged the use of fast-breeding, controllable species, for which most marine organisms did not qualify as well as *Drosophila*. Columbia geneticist Donald Lancefield recalled that he first arrived in Woods Hole to become assistant to Charles W. Metz, (who moved from graduate school under Morgan at Columbia to Cold Spring Harbor and on to the University of Pennsylvania). Metz was then working on *Drosophila* but not on the most popular *D. melanogaster*. He gave still another species, *D. obscura*, to Lancefield as his pet subject thereafter. Morgan and others also brought teams of fly researchers with them each summer, so the MBL had its fruitful fruitfly era, to be followed later by a period dominated by squid.

The early 1960s brought intense neurophysiological work on the highly visible and relatively simple giant squid axon. As supply department head John Valois explains, this is because "two very thick nerve fibers, or axons, run down the sides of the squid's mantle. These fibers coordinate a system of muscles that enables the animal to shoot water through a siphon under its beak, propelling it through the water like a torpedo. But it's the axons themselves, not the squid's behavior, that bring biologists here. Fifty years ago J. Z. Young discovered the axons. He was the first to penetrate one with a tiny electrode to measure the electrical transmission of nerve impulses. Much of what we've learned since then about our own nervous system comes from work on this animal."

Valois explains that the demand for squid has escalated from maybe fifty per month in the 1950s to two hundred (up to four hundred) small ones per day. Currently, about one third of the MBL demand for marine animals is for squid, today's martyr to science as sea urchins, frogs, and guinea pigs have been in the past. The collectors devised special traps and other tricks to gather enough squid in the beginning. Recently the national popularity of seafood has helped. People eat so many of the fish that normally eat squid that the squid have increased in population and are easier to collect now. Because squid do not remain alive in traps for more than twenty-four hours, the increased supply and better collecting methods have saved the day. But even the vigilant supply staff still occasionally has trouble meeting the increased demand during peak season. A sort of marine farm to provide cultivated and more controlled specimens, which Whitman and other directors have envisioned since the 1880s, could help to solve the problem and may soon come to fruition.

Beasts of study are chosen for different reasons. One man, who liked peace and quiet, elected to study the cytology of a particular algae that had to be scraped from the rocks around Woods Hole about midnight. Another researcher wandered into the MBL community by accident, because the Massachusetts Department of Entomology sent him to study corn borers. Long-time MBL researcher Sears Crowell says that he initially wanted to study the coelenterates *Tubularia* and their growth patterns. Or maybe *Hydra*, which have enormous regenerative powers. But, he feared, "the smart guys probably knew about all there was to know" about those two types of organisms. He chose other coelenterates instead. He later realized that hosts of questions remained unanswered about the old familiar friends as well.

The earlier tendency to divide up the world so that each person chose an organism and then asked a familiar set of questions of it has changed. Instead, people in this century have increasingly concentrated on the same few organisms because they provide such excellent material for the questions at hand and because the more that is known about these few, the more that can possibly be known about the larger problems for all organisms. These few serve as systems to model general phenomena. With a smaller number of productive organisms, then, researchers began to ask many different questions. Thus the comparisons made and the spirit of community cooperation were different than those that prevailed in the early decades.

Collecting Organisms

Whatever the organism, obtaining specimens is a crucial part of research at the seashore, as elsewhere. In the first year of the MBL, collecting took place with the help of boats and nets on loan from the Fish Commission, because the MBL did not even own any waterfront property. Furthermore, the MBL itself had only two rather insignificant green dories passed on from Annisquam, which could not go very far in this area of swift currents and changing tides. Collecting under the relatively calm nearby docks or near the shore remained the order of the day. Then, in 1890, the MBL added to its menagerie of small dories the thirty-five-foot *Sagitta*, available for longer collecting expeditions. Then they hired a captain to take charge.

Captain John J. Veeder immediately struck everyone as a perfect choice, and he served admirably as collector and captain until he retired years later in 1933. During his term of service, he directed all collecting trips by student groups, organized class picnics, demonstrated time and time again that he knew how to put on a real clambake, and as the record books always say, never lost anybody. In fact, he was a sufficiently good seaman

The crew of the collecting boat Cayadetta, *1925: Jack Goldrick, deckhand; John J. Veeder, captain; William D. Curtis, deckhand; Ellis M. Lewis, chief; Paul A. Conklin, fireman and oiler.* MBL Archives.

Invertebrate class on board the Vigilant *at Woods Hole dock, 1896.*
Photograph by Baldwin Coolidge, courtesy of SPNEA, Boston.

Setting off on a collecting trip, about 1893. MBL Archives.

Tarpaulin Cove, 1896, after lunch (note the ladies' black umbrellas). Photograph by Baldwin Coolidge, courtesy of SPNEA, Boston.

Lighthouse at Tarpaulin Cove, 1896. Courtesy of SPNEA, Boston.

Collecting under docks, circa 1910. Photograph by Gideon S. Dodds, MBL Archives.

that he avoided any serious troubles and on many occasions rescued stranded or swamped tourists who lacked his experience or good sense.

Students in the courses those early years generally did their own collecting. They went out once or twice a week to dredge and gather the vast variety of life forms to take back to the lab. Until recently the introductory invertebrate course adopted a phylogenetic approach, proceeding through the phyla while studying as many species as possible in each. As the classes grew larger, they ordered a great variety of specimens from the supply department to supplement their own smaller collections. Some students recorded that they were expected to eat what they collected and studied, but that dictum probably found greater compliance for lobsters and crabs than for starfish or flatworms.

With time, researchers as well as some of the courses turned increasingly to the collecting crews for their materials. Veeder served as official collector and helped to provide transportation or to gather species that proved more difficult to find or that localized further away, though the service remained limited at first. For several years it was possible to purchase materials from a competitor, though. "Colonel" F. B. Wamsley learned his trade while working for the MBL, then briefly set up a winter supply business for himself. Early mornings might find the two competitors out scouring the choice locations. The MBL staff of collectors tried hard to

When the dredge comes up. "Dr." Spenser explains things, circa 1909. Photograph by Gideon S. Dodds. MBL Archives.

keep ahead of Wamsley and generally succeeded. Then the MBL hired the Colonel, a South Carolina school superintendent who loved escaping to the northern seashore for his summers. At the MBL he wore his grubbiest slacks and just kept chopping off the bottoms as they ripped and frayed. He was a master at locating and preparing a variety of species, and many people recall his sitting and repairing nets for the next outing.

Despite the augmented staff, the old MBL was never like the Naples Zoological Station, where a researcher placed his order and received his animals shortly after. Gradually, the collecting group and the supply department did develop, so that they supplied most of the laboratory workers' special needs. And eventually they began to take on more and more outside orders as well. In 1896, for the first time, the MBL reportedly shipped about $125 worth of specimens to Williams College to inaugurate the winter sales, which escalated thereafter.

Under Lillie's directorship, the MBL entered into business with the General Biological Supply House. In 1913 a graduate student at the University of Chicago, where Lillie headed the program, saw a biological market and began to fill it. Morris Wells first mailed out forms to biology teachers the next year. He processed his few orders in his parents' basement, provided earthworms or frogs or fetal pigs and brains from Chicago slaughterhouses, for example. Then with a Ph.D. and eventually an assistant professorship at Chicago, he expanded his business. In 1918 he incorporated, with the help of Lillie's father-in-law, Charles R. Crane, who then bought 51 percent of the stock. Crane presented this stock as a gift worth $18,000 to the MBL. The stock value kept expanding so that forty years later

"Colonel" Walmsley and his nets. MBL Archives.

Mending nets in front of Old Main. MBL Archives.

it was worth well over a half-million dollars, with dividends that also exceeded that amount.

Yet this business relationship was not an unalloyed joy. It meant that the MBL collectors spent time preparing things like thousands of pickled starfish and dogfish and less time in gathering specimens for classes and investigators. Great business, maybe, but problematic science. Then the emphasis shifted back toward priority for MBL scientists, and finally the MBL sold its stock to obtain additional funding for building projects.

William Procter with skiff at Great Harbor, 1923. Photograph by Alfred F. Huettner, MBL Archives.

Captain John Veeder in the wheelhouse of the Cayadetta, *1923.* Photograph by Alfred F. Huettner, MBL Archives.

Collecting. MBL Archives.

Generally, supply directors are happy about the reduced demand, as they want to avoid overusing the resources. Veeder, for example, was always careful to avoid overcollecting and threatening the balance of his region. He preferred to collect sea urchins, for example, at the water's edges. He felt that supplies were abundant there, whereas if he disturbed the sea bed, the population might not be sufficient to ensure fertilization of the eggs and

Seining. MBL Archives.

MBL group pulling in the seine with a catboat. The picture appears to have been taken at Quicks Hole, with Pasque and Naushon in the background. MBL Archives.

might fail to replenish itself. John Valois as current supply director has followed a similar ethical dictum.

In those early years, the "second-hand steam yacht," the *Sagitta*, went out twice a week with students on board. One experienced collector on the *Sagitta* reported that the ship was not perfectly designed for the demands. There was little deck room for standing or walking, and the boat rolled very badly even in the rather calm protected waters near Woods Hole. The *Vigilant*, which arrived in 1896, proved much more popular, even if not a perfect ship. This two-masted sailing boat was purchased from a Portuguese swordfisherman and had an interesting shape, pointed at both ends in a manner not typical of American ships. As a result, students called it *Amphioxus*, because it looked so much like the long, thin *Brachiostoma* of that name. On collecting trips, the group of students and instructors gathered on board and nearly filled the decks. They began with the *Sagitta* towing the *Vigilant* off the shore. "Those collecting trips," Winterton Curtis reported, "to dig for worms at Hadley Harbor, collect along the shore at Tarpaulin Cove, dredge off Nobska, angle for urchins and starfish in the Sound, wade the flats at North Falmouth, or new blue crabs at Waquoit were the highlights of that summer." Everyone hoped that the wind would be up and the ship could sail confidently all the way home without the slow and tedious steam help.

"Doc" Hilton extracting flatworms from a conch, circa 1952. MBL Archives.

George Gray with a shark.
Photograph by Alfred F. Huettner, MBL Archives.

James P. McGinnis preparing a dogfish. MBL Archives.

At other times, the students would go out in small rowboats. Curtis recalls that when he served on the staff of collectors, Gertrude Stein attended one of the MBL summer sessions. She had been enrolled as a medical student at Johns Hopkins but had already decided against a medical career. She did not make a particularly flattering impression to Curtis: "For us that summer she was just a big, fat girl waddling around the laboratory and hoisting herself in and out of the row boats on collecting trips."

The supply department followed the Mess in hiring students to help with the increasing work load, as the MBL expanded its extramural specimen supply business. Most students quickly learned to wait on the tables. Learning to sail and to collect marine organisms—and all the other little tricks of the trade—took longer. One young man recalled his arrival at the MBL in 1914. His faculty advisor in Iowa had urged him to go to the MBL for the summer. When he responded that he had to work during the summers in order to attend school during the academic year, he was advised to write and ask for a collecting job. He did, and he received a job. He had to learn a lot, and once lay awake all night in his Candle House bunk practicing the proper knots for making a running boat fast after a particularly embarrassing day. The captain's commands earlier in the day to "take a line" and

On board the Vigilant, *1895.* Photograph by Baldwin Coolidge, courtesy of SPNEA, Boston.

Collecting trip at Quisset, 1897. Photograph by Baldwin Coolidge, courtesy of SPNEA, Boston.

Collecting methods, revolutionized by the introduction of scuba diving. MBL Archives.

"make fast" had rather escaped his landlubber vocabulary. He said he knew that a line was the shortest distance between two points and that he had better do something quickly, but it took him a while to make any further sense out of such phrases. He did catch on eventually and returned to the MBL year upon year thereafter, much later even arriving with the help of more substantial funding from the National Science Foundation.

Others joined the department with greater expertise. Sears Crowell had grown up nearby and often visited his grandparents in Woods Hole. At sixteen, he hired on as "specimen boy," the one who carried the specimens around in buckets to the laboratories. Then he advanced to driving the rubbish trucks to the dump. Then finally he began actually collecting. One of the jobs involved working with Wamsley to preserve specimens from a pond next to the Martha's Vineyard golf course and walking near the greens; the golfers never did figure out what was going on with those strange men carrying buckets. Crowell gained experience and appreciation for the collecting process, which evolved with time, becoming more efficient and requiring fewer collectors as scuba diving replaced more random net dredging for some organisms.

John Valois, when he took over as supply director, was accused by his predecessor of spying on him and of following him around. True, but he had to. Each collector had his own tricks in the game and kept them secret.

They did not keep records of their choice sites and went to some trouble to keep them unknown. But if Valois was to take over, he wanted to know where to find things. So he resorted to such tactics as recording the mileage of trips to collect particularly difficult organisms in order to guess just where the collecting must have taken place. He wanted to learn the system and to bring order and organization into the complex collection enterprise.

Sometimes the collectors went out in search of an unusual catch. One year, before he became assistant director at the MBL, Ulric Dahlgren agreed to deliver a live shark to the New York City Aquarium in Battery Park. The shark expedition began successfully. The team did catch the desired creature, then crated it up and began to tow the crate along behind the *Vigilant*. They made it back to the MBL and showed off their prize. But as they began the long trip to New York, a squall came up. Given the *Vigilant*'s imperfect sailing capabilities, both boat and crate tossed in the waves, and the crate broke loose. Look as they would the next day, the crew could not find it. Only later did they learn, from a friend in New York who had seen a newspaper report, that a shark in a crate had washed up on Long Island and had died shortly thereafter.

Students went on more normal collecting trips and on the annual picnic into this century, with Veeder in control. The good captain protected

John Valois with visitors, Ronald and Margaret Hicks, the Lord Mayor of Falmouth, England, and his wife, 1986.
MBL Archives.

Laboratory glassware. Today's glassware comes from the Chem Room, but in earlier days the MBL had full-time glassblowers, including Jim Graham. MBL Archives.

his charges, including his ships, and would not set out when the weather threatened; with him in charge there were no more exotic whaling stories. His policy was always to leave on time, not waiting for stragglers, especially if the tides or weather were turning. Once even the supervisor of the department, George Gray, was left behind on Naushon Island. He caught a ferry back later, but he was not pleased. Another time a wealthy student did get the better of Veeder's schedule. He was partying with friends and missed the *Cayadetta*'s departure. His friends flew up in a float plane and dropped down right in front of the boat. Veeder had to take him on board. Otherwise, the captain ruled.

With time, more demands on the collecting crew and the increasing size of the MBL population curtailed most such participatory collecting jaunts, and eventually put an end to annual class picnics on nearby islands. Probably the Coast Guard would not quite approve those days of the nonchalant approach to sailing trips anyway; only rarely did anyone use a life vest of any sort. Many long-time Woods Hole residents have their own boats now, often small sailboats or motorboats, in this time when scientists make a little more money than when Whitman urged that no one should become an academic biologist if he expected to make a reasonable living at it. Money does not attract workers to science, of course. As Szent-Györgyi suggested, it has to be a love of science that drives someone to become a successful scientist. That, he says, and an average, but not necessarily above-average, intelligence. What matters is that "you think about it, that you love it, that you live in it, and that's your life."

NOTES

Shinya Inoué's scientific work is discussed in "Exploring the Universe of the Cell," *MBL Science* (Summer 1986) Vol. 2, No. 2: 2–7. Examples of current science come especially from J. P. Trinkaus and Robert Barlow, during personal interviews and through notes, summer 1987.

On early marine work see *American Zoologist* (1988) Vol. 28, especially Jane Maienschein's "Why Do Research at the Seashore?"

On American biologists in Naples, see Jane Maienschein, "First Impressions: American Biologists at Naples," *Biological Bulletin* 168 Suppl.: 187–191; and Philip Pauly, "American Biologists in Wilhelmian Germany: Another Look at the Innocents Abroad," unpublished paper presented at the History of Science Society meeting, 1984. See also Whitman, "The Advantages of Study at the Naples Zoölogical Station," (1883) *Science* 2: 93–97.

Whitman's manual was *Methods of Research in Microscopical Anatomy and Embryology* (Boston: S. E. Cassino, 1885). For more recent publications see Ernest Everett Just, *Basic Methods for Experiments on Eggs of Marine Animals* (Philadelphia: Blakiston, 1939); Donald P. Costello and Catherine Henley, *Methods for Obtaining and Handling Marine Eggs and Embryos* (Woods Hole: Marine Biological Laboratory, 1971); and W. D. Russell-Hunter, *A Biology of Lower Invertebrates* and *A Biology of Higher Invertebrates* (New York: Macmillan, 1968 and 1969). E. B. Wilson's classic work is *The Cell in Development and Inheritance* (New York: Macmillan, 1896, 2nd ed. 1900, 3rd ed., 1928). Keith Benson, "The Naples Stazione Zoologica and its Impact on the Emergence of Marine Biology," *Journal of the History of Biology* (1988) Vol. 21: 331–341, discusses the importance of techniques.

On Lillie's concerns about the MBL, see James Ebert, "Cell Interactions: The Roots of a Century of Research," *Biological Bulletin* (1985) 168: 83.

On Valois's comments on squid as well as the problems of maintaining the seawater pumps, see James Shreeve, "A World Center of Basic Biology Celebrates a First Century by the Sea," *Smithsonian* (June 1988): 90–103 and the full manuscript, which is longer.

On the role of Zeiss: "Carl Zeiss and the MBL," *Collecting Net* (July 1987) 5: 8–9.

Robert Barlow discussed his scientific work in a personal interview in 1987.

Epigenesis-preformation and nuclear-cytoplasmic or internal-external roles in development were hotly discussed issues, as revealed by the *Biological Lectures*. See Jane Maienschein, editor, *Defining Biology* (Cambridge: Harvard University Press, 1986) for reprints of some of those key essays, including William Morton Wheeler's overview of epigenesis and preformation discussions. The classic discussion of such issues appears in Jane Oppenheimer's *Essays in the History of Embryology and Biology* (Cambridge: MIT Press, 1967), and Philip Pauly's *Controlling Life* (Notes, Chapter 3) has very helpful sections as well. The centrality of mechanism-vitalism debates is considered by W. C. Allee, Allee Correspondence 1905–1938, Ed Banks Collection, University of Illinois at Urbana. Thanks to Gregg Mitman for bringing this latter item to my attention. Pauly also discusses the work and ideas of Whitman and Loeb in an unpublished lecture, MBL Archives.

On Loeb's view of collecting and Just's reaction, see Manning's biography of Just, *Black Apollo of Science* (New York: Oxford, 1983).

In personal discussions, John Valois and Sears Crowell have provided particularly useful insights into changes in the types of organisms studied and the types of questions asked over recent decades, as have several of the interviews, Historical Collection. Also, Donald Lancefield discusses early choices of organisms, Historical Collection.

MBL annual reports discuss the General Biological Supply Company, as does Lillie's history (Notes, Chapter 1).

On Dahlgren and the shark, Conklin, "M.B.L. Stories," pp. 127–128 (Notes, Chapter 2): "The Shark Story," pp. 127–128. See also Curtis, "Good Old Summer Times," pp. 4–5 (Notes, Chapter 3).

A number of people recall Wamsley, also mentioned in Lillie's history (Notes, Chapter 1), and Curtis's "Good Old Summer Times," pp. 3–4 (Notes, Chapter 3).

On Whitman and journal publication, see Ernest J. Dornfeld, "The Allis Lake Laboratory," *Marquette Medical Review* (1956) 21: 115–144.

James Graham and his glassblowing appear in *Collecting Net* stories.

7

Out of the Lab

Square dancing at the MBL Club. MBL Archives.

Woods Hole Winter.
MBL Archives.

ONE DAY TWO MEN decided to canoe around Martha's Vineyard—all the way around. A look at the map will illustrate what a distance that is, but equally important is the fact that one side of the island is covered with tall cliffs and offers absolutely no place to land. The fellows succeeded, just as a neurophysiologist recently succeeded in swimming the roughly ten miles to New Bedford one day. The MBL community plays as intensely as it works. The energy and camaraderie carry over from labs to seaside activities.

Boating Trips and Beach Parties

For a while in the early decades of the century, the young men and women spent a great deal of time canoeing around the harbor or around the Elizabeth Islands. Canoeing was the big fad at the time. They would paddle

Canoeing at Hadley Harbor. Photograph by Alfred F. Huettner, MBL Archives.

Canoeing: a sport so popular that in 1926 The Woods Hole Index *devoted a chapter to it.* MBL Archives.

over to an island, then walk to the far end to have a picnic, Isabel Conklin reports. That was all tame and ordinary enough, though sometimes thrilling if the weather got a bit rough. One time, at a party for men only, three canoes headed back in the heavy fog to return to the MBL. After a few close calls and some considerable time heading out blindly into the bay, the group found the shore and resigned themselves to continuing the long way, closely hugging the shore.

Alfred H. Sturtevant and Alfred F. Huettner standing by their canoe, 1923. Photograph by Alfred F. Huettner, MBL Archives.

Beach party.
MBL Archives.

The MBL "boys" would also make a game of swimming across the little space of water, the "gutter," between the tip of Penzance Point and Devil's Foot Island. Just a short space, it seems. But when the tide is changing, the current rips through that narrow space and makes swimming a challenge. Boating through the hole proves even more challenging, and a young man of the 1920s would occasionally try to get his "girl" to try running the hole with him to see if she trusted him well enough.

Beach parties and boating trips have, of course, remained common among the group of young people in Woods Hole. An announcement posted in the post office invited the community to Mrs. Crane's annual birthday picnic on Juniper Point throughout the early decades of the century. One guest recalled the memorable event, complete with a choir from the Russian Cathedral in New York and sometimes a balalaika performance or a dance by a popular summer dance school. The MBL group also gathered for Mrs. Lillie's picnic trip to Tarpaulin Cove. As the MBL group grew too large to fit all together on one annual outing, the classes began to have their own separate picnics; then other informal groups formed as well. As the original researchers became established and bought their own houses in town, and as they returned year after year for their family's summer vacation, the children grew up. Many established close ties with each other and with the town, so that some have decided to remain in Woods Hole, or to retire here, or to visit as often as they can with their own children or grandchildren. Even those who have not become scientists themselves have often returned for the atmosphere here.

MBL picnic. MBL Archives.

Early MBL beach attire. Though fashions have changed, the essence of MBL gatherings at the beach remains the same. MBL Archives.

Marriages

Quite a few families have actually begun in Woods Hole. Wilson reportedly introduced Morgan to Bryn Mawr undergraduate student Lilian Vaughan Sampson, whom Morgan later married. Conklin met his wife while she was a student in one of the MBL courses, while others met at the Mess or while waiting tables, for example. Conklin noted that he had begun to generate a list of MBL marriages until he realized that it was longer than a "Catalogue of the Ships of Homer." If "marriages are made in Heaven," he remarked, then "there is certainly a large branch office in Woods Hole." Furthermore, he enthused, such marriages were eugenic and likely to last since the participants shared interests and companionship very generally.

One Woods Hole proposal had unusual complications. During the first war, the Fisheries area was closed off by a marine guard. Anyone who worked there could only be reached by written message. Accordingly, an MBL scientist asked for a messenger to deliver a note to a woman at the Fisheries. The note, it seems, contained an invitation to go canoeing that afternoon, at which time the scientist proposed.

The war did not interfere with the important business of life, though sometimes science did. It was not always easy being an MBL spouse or child. Summer resort housing did not offer as many conveniences as at home, the nonscientist spouse (usually the wife) often felt left out of the interminably scientific conversations, and it was hard having a husband who was away working in his lab at all hours. New brides found the circumstances particularly unsettling, though they either adapted and fell in with the different style or else left.

The Beach

As families have learned to relax and enjoy, MBL folk have taken an ever-more-active interest in the beach. A sunny afternoon finds the MBL beach full of teenagers and older sunbathers of all sort, though these *are* busy scientists so that many choose early mornings and late evenings for their swimming. One non-MBL Woods Hole resident complains that she does not like the MBL beach. People there talk in funny accents about things that are not comprehensible, she laments. And she is right; this beach on Buzzards Bay, which was donated in 1936 by Edward Meigs and enlarged with an addition by Oliver Strong in 1940, is unusual as a beach. No penny arcades; no cotton candy; no hotdogs here. This is the MBL beach.

As Lewis Thomas has so ably written, it is a special place. Scientists do not set aside their research work and forget about it for a few hours of fun on the water's edge. They take their work, in their heads, in their books, or scribbled on a bit of paper, with them. Many a dilemma has been resolved

MBL beach. Photograph by Alicia Hills. MBL Archives.

or an attack set up through a conversation at the beach. Wonderful solutions to troubling problems find their expression in the MBL sand: whether a technical trick for getting something done or planning some exciting cross-disciplinary symposium that would not occur elsewhere, where people would not even have opportunities to talk to each other.

Some people go in the water, and when the tide is high and the wind calm many addicts can be found moving back and forth, back and forth, swimming those laps to which they have become so accustomed in their civilized university pools back home. Then there is the time when Szent-Györgyi swam from his place on Penzance Point over to Juniper Point, some considerable distance. A friend had promised him breakfast any time he wanted to swim over. Unfortunately, his friends were out of town, and the woman renting the place was not at all amused. She would certainly *not* provide any breakfast and would barely let the dripping Szent-Györgyi use the telephone to call a taxi.

At the MBL beach, when the Portuguese man-of-war or some other tyrant makes a visit, some people head out of the water; others rush for their collecting nets and scoop up samples to show the children or to investigate themselves. The other beaches are altogether another matter. There people are more likely to talk about normal things. They may bring their picnics or their wind-surfing equipment and settle in for a relaxing afternoon reading a spy novel. Some fishermen cast off from the rocks below the lighthouse. The difference in atmosphere represents more than just the different sides of the spit of land on which Woods Hole sits.

Drama

Beyond the beach, the MBL crowd has become involved at various times with other outlets for recreational energies. One year in mid-century a group of young scientists and friends agreed to help raise money for the Woods Hole library and the MBL Club at the same time. They agreed to perform a vaudeville type skit entitled "Tea for Several." They had a magician, and they sang songs for the other acts. But the play was the thing. They had used local people as models for their characters, and some complained. A few also complained that the show was vulgar—thereby demonstrating that they had probably never seen a vaudeville act and that the standards of the Laboratory and the town dowagers were not always in perfect sympathy.

Another performance much earlier had had a different ending. History suggests that Englishman Bartholemew Gosnold had first arrived on Cape Cod and the Elizabeth Islands in 1602. Nobody noticed in 1902 that it was

J. K. P. Purdum as Gosnold in 1907 before the fall. Sumner family papers, MBL Archives.

time to commemorate that event of three hundred years before. When they finally did take note a few years later, they decided not to miss the opportunity altogether. Woods Hole would enact a landing by Gosnold, with Fisheries fish culturist J. Purdum playing the part of the hero. As George Howard Parker told the story, a group rowed the serious Purdum to shore in an old boat, properly decorated for the special occasion. "A plank was put out on which he was to walk to the beach to greet a second body of villagers dressed as Indians. Just as Mr. Purdum stepped onto the plank, a volley was fired unexpectedly from the stern of the craft." As a result, instead of landing in appropriate style on the beach, Purdum fell unceremoniously into the sea. Nonetheless, the hero "waded out of Great Harbor and he completed his part of the performance without cracking a smile."

E. B. Wilson making music with his daughter Nancy. Photographs by Alfred F. Huettner, MBL Archives.

Music

Music, including classical music, has always played another central role in Woods Hole and in MBL society. In those early years, Wilson was a first-rate cellist. Indeed, he had made a hit when he visited the Naples Zoological Station as much because he knew and loved music as because of his outstanding cytological research. His daughter became a professional cellist, but the Wilsons were really just the first of a long series of MBL performers. Frank's younger brother, Ralph Lillie, was a marvelous pianist and often played for friends. And Al Romer and Ricky Harrison (Ross Harrison's oldest son) often shared songs that they had learned, respec-

Jelle Atema, musician and scientist, 1986. MBL Archives.

Fifties student making quiet music in the laboratory while others sit at lab tables. MBL Archives.

tively, while tramping through England or while visiting German relatives. Albert Szent-Györgyi allowed that he was not particularly gifted as a musician himself but that he had grown up in a musical family. His mother reportedly decided against a career as an opera singer, on the advice of Gustav Mahler, who told her that she did not have the necessary physique for a singer.

Today, the piano in Swope rarely sits idle in the evenings, and people sit outside on the lawn listening to the music wafting out the windows. A recent celebration of the work of Woods Hole artist Alix Robinson included a series of concerts, for which as many people may have sat outside on the grass as inside in the formal seats. The regular Sunday evening concerts always attract a full crowd as well. So many MBL people have said that the musical offerings in the little town of Woods Hole remain one of the major drawing points. The official centennial schedule included a performance by premier flutist Jean Pierre Rampal of composer Ezra Laderman's "The MBL Suite," composed for the occasion. Music gives a cosmopolitan and educated air to the setting, although the MBL music sometimes clashes with that blasting from the large radios of the vacationers waiting for the ferry.

Hurdy-gurdy music can clash too. One story goes that Conklin, who had no musical ear at all but who knew quite well that Wilson did, hired a hurdy-gurdy man to play outside the labs. The man was playing down the street, complete with monkey and tin cup. Conklin then hired him to stand under Wilson's window and keep playing his grating music. "Don't stop, even if the man inside says to," Conklin insisted; "he just doesn't want to pay. I have already paid, so keep playing."

The MBL Club

On a slightly different note, the MBL Club and Woods Hole itself also include a healthy dose of sea chanties, and folk singing of various sorts reflecting the changing times. The MBL Club also offers dancing for the community, sometimes with general events and sometimes with boisterous and popular teen dances. In the early years, dancing more often came in the form of a visit from Josephine the Bear, since before 1918 the people at the MBL did not dance much. Only when World War I brought the first temporary naval base to Woods Hole did dancing become a regular feature. Sailors liked to dance, but more importantly, with the sailors came a victrola, and that victrola attracted dancers. World War II had an effect as well, since gas rationing kept people at home looking for something to do. The uniformed servicemen attracted attention to light-hearted dancing, even as the MBL rented out buildings to the war effort and as the planes flew overhead. Then as society became more liberal, and as the MBL Club provided a familiar

MBL Club mixer. MBL Archives.

and comfortable place to relax, people began to use it more often. In recent decades, the old firehouse down the street has hosted occasional ballet classes or folk dancing sessions as well.

The MBL Club opened in July 1914 in the building of the old yacht club. As with so much of the MBL, the building was a gift from the indispensable

Josephine and fans. C. D. Whitman stands at the top of the stairs on the right. MBL Archives.

Charles R. Crane, who felt that the young men and women needed a place to meet, as the laboratory crowd and the wealthy summer crowd did not mix much then. There was relatively little social life for the students and young investigators who were not married and did not have houses. The MBL Club promised to remedy that.

Actually, the MBL Club should have opened in 1913 with the gift of the building, but researchers needed the space for laboratories until the next year when the Crane Building was completed. Very soon thereafter, the directors introduced denatured alcohol for their "teas." Magazines made an appearance a few years later. At first the Club offered a place for a few workers to gather for a smoke in the evening, but as people's living quarters grew more scattered they turned to the quiet, relaxing setting of the Club to gather and talk. They might discuss a recent book or movie for a while, but conversation inevitably drifted back to the latest scientific idea or technique. While the director, Merkel H. Jacobs, in 1930 officially applauded the MBL as a place of individualism but cooperation and a "center for healthy critical interchange and stimulating contacts of individuals," others seem to have partaken of the lighter attractions there.

The MBL has continued to provide a place for people to gather informally. At first everyone at the MBL could gather on the front porch to talk or sing and smoke. Later they could still all fit into the boat for a picnic or together in the Mess. As time progressed, more and more people have come, of course. They have also come at irregular times, with staggered schedules so that everyone is not new all at the same time. "Other people all seem to know each other, but I don't," someone complained recently. Some people are in more of a hurry when they are here, trying to squeeze in visits to old friends with time in the library or collaboration with a colleague in the lab. Some noted the speeded-up pace and the crowded calendars as the centennial year approached. People live farther away so that not everyone is in easy walking distance of the main buildings anymore, which produces greater anonymity than before. Yet the MBL Club has nonetheless continued to provide a setting for social interaction during all that change and hurry, whether on Sunday morning over newspapers or with dancing and singing.

Singing

Singing became a favorite evening activity right away, and the founders intended singing to serve as a central focus of the Children's Science School. It did, but no longer does everyone know the same songs. There is no common repertoire so widely shared as in earlier times, so singing at the

MBL Club involves a good deal of learning the words as well as the notes. No longer do groups join in enthusiastic chorus to sing to the tune of "Tipperary":

A fish-like thing appeared among the Annelids one day.
It hadn't any parapods or setae to display.
It hadn't any eyes or jaws or central nervous cord
But it had a lot of gill slits and it had a notochord!

It's a long way from Amphioxus
It's a long way to us.
It's a long way from Amphioxus
To the meanest human cuss
It's good-bye fins and gill slits,
Welcome skin and hair
It's a long way from Amphioxus
But we came from there.

. . .

My notochord shall grow into a chain of vertebrae
As fins my metapleural folds shall agitate the sea,
This tiny dorsal nervous tube shall form a mighty brain
AND THE VERTEBRATES SHALL DOMINATE THE ANIMAL DOMAIN.

Students remember such songs from the MBL, but also from other places, as the MBL community took its lessons elsewhere. One former student recalls studying biology at Harvard from Museum of Comparative Zoology director Alfred Romer. He came in, sat on the desk, and sang the whole song, with its various verses, about the long rise of vertebrates from the lowly Amphioxus. It worked; many years later she still remembers the lesson.

Another favorite went to the tune of "Sweet Marie":

It's a question to my mind, sweet Marie,
What in annelids you find, sweet Marie,
Can you number and confirm all the segments of a worm?
Do you know the mesoderm, sweet Marie?

CHORUS: *Sweet Marie, look and see,*
Look and see, sweet Marie,
Tell me what without the lens you can see.
Do you think you'd better try
With your own unaided eye
To distinguish nuclei, sweet Marie? etc.

Walking, Biking, and Running

In addition to these basic Woods Hole pleasures, walking takes its place as a favorite. Earlier generations may have hitched up their skirts and doffed their coats and ties to head for a relaxing break at Nobska Point or farther north. Or they may have headed up Mowing Hill toward what is now the golf course. From there, you could see all of Woods Hole without the trees to interfere. A few took off on long Sunday treks, some wandering twenty or more miles to explore the Cape, covering ground now occupied largely by roads and development. Of course, now people are more likely to drive to a public park or somewhere else "official" to do their hiking, often forgetting about their own back yards.

Walking is more dangerous today along the streets where cars drive faster and there are no shoulders or only healthy fields of poison ivy to dive into. Yet the bike path provides an easy walk into Falmouth, winding along the beach and through humid dense greenery. A visit to the wildlife sanctuary at Quissett Harbor is also well worth the walk, especially at sunset as

Mary Huettner at Nobska Light, 1920. Photograph by Alfred F. Huettner, MBL Archives.

the elevated rocks there provide one of the truly spectacular and uninterrupted views toward the west. Walking home along the narrow unlit road at night takes some courage and agility but can be executed successfully.

Bicycling and running have largely replaced walking for many, of course. Isabel Morgan Mountain conjures the image of six Morgans streaming down the hill on their six bicycles in the early part of the century. And over the years many people have encountered the question, what to do with those bicycles during the year when they return to their other home. At various times bicycles have probably spent the winter stashed in most Woods Hole garages and storage sheds.

Although there do not seem to be quite so many sweat-soaked runners as there were a few years ago, the streets and bike path do boast a healthy number. Many are training for the famous Falmouth Road Race, which occurs every August. In its first years the race occurred at noon on a Sunday. But the increasing thousands of competitors and doctors' warnings about the effects of running in the beastly heat and humidity, which occasionally do occur, moved the time to earlier in the day. The race, which begins in Woods Hole, winds past the Nobska lighthouse and along the beach, and ends in Falmouth, is quite an event.

Fourth of July, Diving, and Sports

The Fourth of July parade is also an event, sort of. Hardly a fancy parade in the traditional sense, this annual stroll down Main Street is more a meandering than the patriotic exhibition typical of most small towns. Everyone can join in, and many do. People dress up in all sorts of outlandish outfits. Then the parade just peters out: no fancy reviewing stand; no dignitaries to watch and bow or raise their hats appropriately. Just some good old fun reminiscent of a different era, and maybe some watermelon or other refreshments afterward. Sometimes boats also get decked out like they used to.

The Fourth of July used to include diving contests, and Isabel Conklin recalls one particularly exciting year. All year the dock had a lot of fishing and diving going on, with diving practice from the bridge as well. But that Fourth of July one of the women, a very good swimmer, shocked the crowd. She had on a proper bathing dress, of course, complete with the requisite sleeves and ruffles. Then as she prepared to dive, she suddenly pulled off her suit. Underneath she had on body tights, or what were in those days called "Annette Kellermans."

Years later, when a visitor revisited Woods Hole after a long absence, his hosts took him all around the labs, the Mess, and the other various buildings, as well as to the beach. Then he was asked what he found most

The Cayadetta *"dressed" for the Fourth of July and about to set out on a picnic, 1932.*
Donald Zinn Collection, MBL Archives.

different about the MBL. After a minute, he responded, "Well, I tell you my boy, I really think it's the change in the bathing suits that the girls are wearing now."

Other activities take groups out on the fields. Horseshoes was a popular sport in the 1920s especially. Then baseball and softball gained attention, made possible by the generous gift of a town ball field from the Fay family. This undeveloped piece of tucked-away land finds many users in the summer months. Sometimes it has been soccer that has attracted the most enthusiasm; otherwise, frisbee or other games.

Tennis is a necessary part of life for some at MBL, perhaps even required to stay healthy and alive. Not often does one find an empty court on a calm, comfortable summer day in Woods Hole. And the really windy days make for interesting play and even more amusing watching. The tennis courts next to the MBL beach were given to the MBL along with beach property and are operated by the Tennis Club, whose notebooks in the Archives attest to the officers' meticulous accounting. Tennis is big at the MBL, so that developers will probably have trouble removing the courts to put in new lab buildings if they want to—unless they provide alternatives elsewhere. Of course things have changed. With growth and progress, the tennis courts no longer lie right behind the Mess Hall. So not everyone gathers around to watch the interminable after-dinner games or the final matches of the annual tournament as they used to. Not everyone in town even knows who the chief leading competitors are this year.

Poker games are popular, too, though perhaps not on grounds of health. The ongoing games bring people together and, once again, provide an opportunity to discuss science and to bring together that cross-fertilization of ideas and innovations that is the MBL. A biochemist and a neurophysiologist might have some clever ideas about neuromuscular action over a straight flush, for example.

Gardening

Gardening fills the time of many Woods Hole residents, though most summer scientists do not choose to invest the time or the energy in a garden that they miss for so much of the year anyway. Because of the humidity and

E. E. Just throwing horseshoes, flanked by Calvin Bridges and Donald Lancefield. Photograph by Alfred F. Huettner, MBL Archives.

bountiful growth, however, people with land generally have to cope with the plant population to some extent. Pulling weeds, trimming trees, trying to keep back the poison ivy and scratchy wild Cape Cod roses: all this takes time. Those MBL kids, who pleaded to spend the summer in Woods Hole in order to see their friends again and to take courses again at the Science School, may find themselves on enforced garden duty. Many yards reflect the attention given, however grudgingly, and reward the casual walker out for an evening stroll in the balmy air. Sometimes a scientist has embarked on botanical experiments at home. Morgan, for example, bred various plants (such as verbena) as well as mice and occasional birds during his Woods Hole summers.

On a more formal and scientific note, the herb garden next to the bell tower offers a quiet place for sitting and thinking. Designed by landscape architect Dorothea Harrison (one of Ross Harrison's daughters) and kept in shape for a while by her and her friends, the garden receives less attention from visitors these days, but has been well tended by someone. Another Harrison offspring, Richard Harrison, became an artist of a different sort, doing architectural work and drawing for the *New Yorker*, for example. The scientific interest in order and drawing rubs off on the talented children in various ways.

Sailing

After a busy day of walking, singing, visiting gardens and neighbors, and swimming, many people would love to take a sail. If only they could get to know someone important or generous, they could. The Eel Pond is beautiful with its complement of sailboats. Perhaps all that opulence bothers some of the more politically egalitarian scientists, but the recent invasion of those sailboats is not likely to stop for that reason. Of course, smaller boats have been around for a long time. More than one MBL worker has purchased a boat to get around the choice collecting sites well before investing in land or even a car. Each winter some of those small boats find their way to the bottom of the Eel Pond, alongside old-fashioned discarded microtomes that some people used as anchors and other odds and ends. Occasionally the ice and storms sink larger and more expensive boats as well. Spring cleaning has its own meaning at the seashore.

Meals, Movies, and Diversions

After all that activity and a possible sail, anyone would surely be hungry. Outdoor grills give evidence of the swordfish dinners or the occasional clambakes out of doors. Lots of good, relaxed scientific discussions take

Winter at Eel Pond, 1971. MBL Archives.

place over a whiskey sour, a shrimp cocktail, and a slice of fish. The visitor who wants a break from Swope food or who learns that no meals are served there on Sundays may head for the Black Duck Restaurant for breakfast or brunch. There one can sit down to a gigantic meal next to the water and watch the occasional ducks, or gulls, or on a warm summer's day a well-behaved mother skunk and baby as they scurry under the dock. Or one can go to the Fishmonger for an ice cream cone. Or the Food Buoy for a yogurt, soda, and candy bar. Or to the liquor store. There are other, more tourist-oriented places as well. Other times call for a group invasion of the Captain Kidd Bar—with its unique mixture of sailors, tourists, and scientists—for a restoring libation after a particularly tedious or inspiring lecture, or for no reason at all.

After such a busy day, especially the young folk and teenagers may want to take in a movie. If the MBL is not showing one or if it is too familiar, one can head for Falmouth. Groups used to hop the train or take a bus, or even

Main Street, Falmouth, circa 1932. Photograph by Alfred F. Huettner, MBL Archives.

to hire a taxi for twenty-five cents—to see a twenty-five cent movie. Others walked into Falmouth for the big show. Once a group persuaded the absent-minded Robert Chambers to drive them over. Afterwards the group disbanded because some wanted to walk back and others decided to get an ice cream cone first. Chambers got a ride with another group, forgetting his own car. The next day he notified the police that someone had stolen his vehicle. Soon they reported back to him that it was found in front of the movie theater and that he had better move it immediately since it was illegally parked.

Woods Hole does not offer much variety, but it does offer simple diversions. For a number of years after Prohibition, people gathered to watch the Dude train roll in from Boston, for example. It was always amusing to see who had enjoyed the bar car a little too much this time. In Woods Hole, one really does not need much more in the way of entertainment. Scientists are supposed to stay in their labs working, emerging only on occasion for such breaks as mentioned above, are they not? They do at any rate. The fact that some types of scientific work allow breaks may help explain why particular people have chosen the work they have. Fortunately, the library reader can register for a reserved desk (or as a general reader) and can settle into the stacks to work in a place that allows adequate opportunity for staring out the window too. Especially the desks on the top floor: all look out toward water in whichever direction. "Oh, I was just thinking, outlining the next chapter," the daydreaming reader can claim. In contrast, taking even a visual break from most labs back home requires actually leaving, thereby admitting even to oneself that one *is* doing something other than working. At the MBL, though, the pace is intense, with few real breaks in the ongoing pursuit of science.

Life and research go on, even when other people are visiting the beach or sleeping. During the summer, someone is at work somewhere in the MBL at almost any given time of the day or night. That, too, is part of the spirit of the place. Though the researchers and the supporting staff no longer have to get by with quite the meager resources once at their command, they still have to work hard to keep up the elusive spirit of the place that grows out of the camaraderie and the perpetual discussions of science that go on in so many different settings during the day. Admittedly, things are not what they used to be; in many ways they have become even better as the community has expanded.

The Dude Train, 1895. Photograph by Baldwin Coolidge, courtesy of SPNEA, Boston.

NOTES

Isabel Conklin and others in their Historical Collection interviews discuss the importance and popularity of canoeing. On proper canoeing technique for Woods Hole in particular, see E. V. Cowdry, "Canoeing in the Vicinity of Woods Hole," *The Woods Hole Index* (August 1926) No. 2: 1–5.

W. C. Allee, in notes on the MBL in University of Illinois Archives, describes Mrs. Crane's birthday parties.

George Scott, Historical Collection interview, recalls the Purdum and Gosnold adventure.

Many of the Historical Collection interviews refer to Woods Hole as a place where marriages were made, including references to Conklin's comments on the subject.

On music: Wilson's love of music was legendary; other information comes from Historical Collection interviews by Szent-Györgyi and Isabel Conklin in particular.

In a personal interview, Sears Crowell recalled the hurdy-gurdy story, though he cautions that his memory may not have the details just right.

On the MBL Club and the role of singing, many people today attest to the continuing interest, also documented by numerous items in the *Collecting Net* (see informal index, MBL Archives, for specific references). Several informal songbooks exist in the MBL Archives.

On changes in the bathing suits, see the Reznikoff interview, Historical Collection.

A number of those interviewed in the Historical Collection and later attest to the role that movie-going played, while Paul Reznikoff's interview tells the Chambers's stories, Historical Collection.

Road to Penzance, circa 1910. Ida H. Hyde Collection, MBL Archives.

8

Friends and Relatives

Eel Pond: carrying sails out to spritsail. Photograph by Baldwin Coolidge, courtesy of SPNEA, Boston.

View from the roof of Lillie, looking toward the ferry dock. MBL Archives.

THE MBL SPIRIT is not confined to a few acres of MBL property. Rather, because the spirit is characterized by a free and cooperative exchange of ideas and beyond usual boundaries, the long-term development of many ties to other people and other institutions spreads that spirit widely.

Most obviously, thousands of students have spent time here in hundreds of courses. Thousands of instructors have taught those courses. Thousands of investigators have pursued research here joined by a host of workers and visitors and library readers and staff members. The fact that Japan's Emperor Hirohito chose, of all places, to visit the MBL attests to its sphere of influence. People from all over come through for tours of the place they have heard about. Not all the people who have become part of the MBL's 100 years have stayed long; but they nonetheless visit the place and take away something, if only a photograph or T-shirt and a vague sense of what goes on in this place of science, which they then describe to friends back home.

In the 1920s, Theodore Dreiser visited the MBL at the urging of a friend from the Rockefeller Foundation. He found just what his friend had promised: an exceptional place. Expounding in a manner typical of the period, he enthused about what he saw as the unbiased and pure work before him. As he put it, "The patience, earnestness, and, I assume, honesty of these men and women impress me more than anything else I have seen in America." The community was embarked, he thought, on a marvelous and beautiful search, "the most honorable and respectable employment of man," like going into battle or like hunting or exploring a house with locked doors and no keys. "My compliments to the workers of the Marine Biological Laboratory of Woods Hole! A profound and reverent obeisance."

As the century progressed and the "ring of marine stations around the globe" that Naples station director Anton Dohrn once envisioned has gradually appeared, many people have appealed to the MBL for advice and guidance. Naples set the standard at first, with superior facilities and equipment. But at Naples, investigators worked in their wonderfully quiet rooms upstairs, while the public passed through the open aquarium downstairs, paying their admission fees to support the cause of science above. The Naples station did not teach courses or bring the extended coterie of assistants and graduate students that very quickly became a vital part of the MBL. The MBL offered a model (though not always a perfectly successful model) for integrating instructional and research purposes, and other stations have followed. Within the century, the coastal states and most countries of the world that touch on a seacoast, as well as many that do not, have built their own marine stations. Often they have followed the MBL model, seeking to foster that spirit of free interchange and cross-fertilization among students, teachers, and independent researchers. Others have purposefully diverged in whatever ways to develop some other goal.

As these many stations have grown, they have often appealed to the mature and experienced MBL for advice: how does one get the seawater system to work without corroding all the metal fixtures or polluting the environment with toxic metals, for example? How does one procure large numbers of specimens without overfishing the local waters? How does one combine the various desirable functions of a marine laboratory at reasonable costs to the participants, without going too heavily into debt? Some labs have had wonderful benefactors who have given them more solid and long-term financial security than the MBL has so far achieved. Many are run by a single institution such as the excellent Friday Harbor Laboratory at the University of Washington, or the Scripps Institution of Oceanography at the University of California at San Diego. The MBL has resolved time and again to remain independent, as a truly national and even international facility, but at great cost at times. Only a solid endowment will ensure the successful continuation into the second century of the myriad quality programs and of the free atmosphere of open discourse. The MBL must learn from others how to pursue such a goal.

In Woods Hole, the MBL also has a vital host of friends and neighbors. Many of the wealthy land owners, beginning with the influential and generous Fay family, have given money, land, and advice at critical junctures. Charles R. Crane followed, then Carnegie, Rockefeller, and more recently the Grass, John D. and Catherine T. MacArthur, and Andrew W. Mellon foundations, along with many other donors to make the laboratory's continued growth and improvement possible. Individuals from Penzance Point, the Cape, and elsewhere have generously donated time and money to the MBL's well-being. This support is essential, but probably not sufficient without a solid permanent financial base. Even the crucial substantial grants for many researchers from the NIH and NSF cannot cover all the ever-increasing expenses for sophisticated world-class science.

Equally important for the MBL's well-being are the local families who have contributed indispensable employees. Some have begun as maids or collecting crew assistants and have risen to head their departments. Their energetic actions in hurricanes, at the beginning of summers when many people rush in at once wanting everything immediately set up just right, or in other times of need demonstrate their work well beyond the call of duty. Many of these people, some second- or third-generation MBL, believe in the MBL as an important place of science and want to be part of it. So many cases abound—of the specimen boy who rose to become instructor or even trustee, the waiter who became lab director, the dining hall chef whose father ran the earlier food service, or the local youngster made head of his department. Many of these people could have made more money elsewhere, but they have made their own generous contributions to the MBL.

The MBL Associates is a supportive group that has arisen since 1944 out of similar affections for the place. People associated with the MBL but not all as researchers or course instructors have wanted to help. Special projects such as the Futures in Science "FiS kids" program or the organizing of photographs or acquiring of rare books for the Archives and Rare Book Room have attracted the attention of the MBL Associates and other volunteers.

Aside from its many invaluable friends around the world, the MBL also has an important set of close relatives. Without Spencer Baird and the Fish Commission, the MBL Supply Department could not have come into existence as such. It was Baird's enthusiastic encouragement of Hyatt, his arranging for the original land to be given, and his generously supplying the seawater, collecting boats, and much equipment in the early years that made the MBL happen. (Baird actually set up the agreement but died before the MBL opened.) Baird's supporters, who had subscribed to tables for their students to do research at the Fish Commission, did not see the point of encouraging this competition. In particular, Brooks at Johns Hopkins University and Alexander Agassiz at Harvard remained skeptical or even hostile towards the new enterprise. Who needs two labs in Woods Hole, they asked, particularly when one funded by the federal government would be far better? Who cares about teaching students and high school teachers, when research really matters? Baird did recognize the value of the MBL enterprise and without jealousy paved the way for what he realized might become competition for Fish Commission research. Through the twentieth century, the relationship has remained relatively distant because of the Fisheries' government mandate to pursue practical work more than research and teaching of biology. Yet the institutions have continued to cooperate in a variety of ways, loaning each other boats or other equipment, or even scientists at times.

The Coast Guard has remained even more autonomous, because of its very distinct functions. Yet MBL crews and the Coast Guard crews have helped each other in important ways over the years. And MBL scientists have always enjoyed visits to Nobska Lighthouse, run since 1939 by the Coast Guard when it replaced the old Lighthouse Service.

The MBL's relationship with the Woods Hole Oceanographic Institution (WHOI) is the closest. In fact, WHOI is really the MBL's younger sibling. It began when MBL director Lillie and Wickliffe Rose, president of the General Education Board, decided that an east coast oceanographic research center might be in order. The Pacific United States had its Scripps Institution of Oceanography. Since it was not clear exactly what sort of eastern establishment should be developed, the National Academy of Sciences set up a study group. As a result of that group's report and subsequent developments, January 1930 brought the official incorporation of Woods Hole Oceanographic Institution.

Why, Woods Hole people often ask today. Woods Hole is so crowded already and as WHOI has expanded it has had to set up a second, separate campus site, the Quissett Campus. Lillie reported that Woods Hole "which had from the first been regarded as the most likely site, was definitely selected, on account of its geographical advantages and the scientific good will and co-operation assured there." Also the existence of a library was a major attraction. Lillie, as leader of the MBL and inspiration for WHOI, clearly sought to keep both in Woods Hole.

Lillie reported that the MBL Board of Trustees wanted to donate a piece of land to WHOI in the first place. It later turned out that they legally had to be paid for the gift, but that commitment nonetheless set the stage for the two labs to be neighbors. Yet later, when the Oceanographic wanted to expand and buy the MBL's Penzance Garage property, then-director Philip Armstrong reported, the MBL did not agree. Experts and real estate specialists from Washington came to assess the properties and concluded that Woods Hole had "too little land" and "wouldn't develop a well-integrated campus." The Navy took the property anyway by eminent domain, then released the land to WHOI. Such actions, given the very limited amount of land and especially deep waterfront property, temporarily created some hard feelings and stimulated a bit of sibling rivalry. Fortunately any such feeling has largely dissipated.

Some MBL old-timers felt that the Oceanographic had "sort of ruined Woods Hole." It seemed so much more crowded than it had before, with so many autos and so many people. The smaller MBL has sometimes envied the large endowment and the solid financial basis of the grown-up little brother. At other times it has been disdainful. The World War II "booms" heard frequently off the shore broke windows and set people complaining. Even if the experimental explosions were part of the war effort, why here in Woods Hole? they asked. During World War II, WHOI thrived while the MBL shrank. More recently, directors and staff have worked at building stronger cooperation to enhance the whole Woods Hole scientific community.

Crane Laboratory viewed from Eel Pond, 1923. Norman W. Edmund Collection, MBL Archives.

MBL picnic: coming home, early 1890s. Photograph by Baldwin Coolidge, courtesy of SPNEA, Boston.

What the nostalgic criticisms and occasional frustrations have failed to appreciate is the fact that someone would have been there anyway. It is better to have scientists, with many shared purposes and common interests. As the various parties involved now realize, the community is stronger when the diverse groups share activities.

MBL science is relatively small-scale science, done largely by visitors. They have grants, of course, and assistants and equipment. And MBL does have absolutely first-rate work in cell biology, physiology, the neurosciences, and other fields, so that it is one of the world leaders in those areas. Yet overall the MBL remains on a much smaller scale than the Oceanographic. WHOI was set up with intended connections to the various armed forces and other government and private groups. It employs a much higher percentage of permanent year-round researchers and staff. It receives large government contracts for a wide variety of types of exploratory work.

When the *Titanic* came to town in 1986, in the form of wonderful photos, folios, artifacts, and lots of stories, the MBL welcomed the WHOI team back and congratulated the conquering heros. A few may have felt a little irritated by all the attention ("that can't be real science"), perhaps

MBL says Bravo! Photograph by M. Rioux, MBL Archives.

briefly wishing for some of the limelight themselves before they settled enthusiastically back to their own work. Brothers and sisters are like that: fond of each other, annoyed with each other, and usually, deep down, supportive and fiercely loyal to each other. Especially in winter, when crowds thin and paces slow down, cooperation in the scientific enterprise prevails. People from WHOI and the MBL attend brown bag lunches together. Some serve as trustees and on advisory groups for both institutions. There seems to be no point to being competitive in this little town at the end of Cape Cod when the wind is whipping through and the ice collects on boats and water.

As a result of the generally supportive scientific atmosphere, the U. S. Geological Survey has set up offices here, as has the Sea Education Association. The National Academy of Sciences has established a conference center in a beautiful old house on the water nearby.

The Fisheries, WHOI, USGS, SEA, and MBL scientists work together in various ways, expanding the spirit of scientific activity and making Woods Hole a special place for science. The Fisheries and Aquarium serve more direct commercial and public purposes, WHOI carries out government and private research, while the MBL provides for independent scientific research and instruction at all levels. Here is specialization and cooperation: just what Whitman insisted should provide the basis for effective science.

Eel Pond, looking toward the golf course, 1895. Photograph by Baldwin Coolidge, courtesy of SPNEA, Boston.

NOTES

MBL annual reports provide information about the course members over the years, both names and numbers of participants.

Theodore Dreiser's visit was discussed in the *Collecting Net* (July 21, 1928).

Ralph Dexter on marine laboratories, (Notes, Preface) discusses the list of current facilities.

On other activities and groups in Woods Hole, see the recent publications: Mary Lou Smith, editor, *Woods Hole Reflections* (Historical Collection, 1983) and Mary Lou Smith, editor, *The Book of Falmouth, A Tricentennial Celebration, 1686–1986* (Falmouth Historical Commission, 1986).

Paul Galtsoff, *The Story of the Bureau of Commercial Fisheries Biological Laboratory, Woods Hole, Massachusetts* (Washington, DC: U.S. Department of Interior Circular No. 145, 1962). Curtis, "Good Old Summer Time," (Notes, Chapter 3) recalls the sharing of equipment between the USFC and MBL.

Lillie's history (Notes, Chapter 1) still provides the best look at the early years of WHOI, quotation on p. 180. Editors Mary Sears and Daniel Merriman's *Oceanography: The Past* (New York: Springer-Verlag, 1980) also provides information about the Woods Hole Oceanographic Institution and insight into the history of the Woods Hole establishments, including WHOI, as well as about the history of oceanography more generally.

Armstrong discussed his reactions in his interview, Historical Collection.

Yacht visiting Woods Hole. MBL Archives.

Thanks for the Memories

Sources and Acknowledgements

Discussion in this book is based primarily on materials found in the MBL Archives, mostly written, documented sources. Often the same event or fact appears in more than one place. Some stories come, with permission, from tapes of interviews made by the Woods Hole Historical Collection during the 1970s and 1980s. And a few items emerged from interviews that took place at the MBL.

I chose not to try to acquire new material by interviewing everyone in the MBL community, but instead spoke only with a few of those who have been around for many years in order to verify information or to fill in details. This project, then, is autobiographical in that it is based on information chronicled within the MBL records. It is not an attempt to provide a more detached or analytic biographical treatment. I hope this book will facilitate other such projects.

In particular, I wish to thank the Woods Hole Historical Collections for generously sharing their transcripts of taped interviews with MBL scientists and staff members. Interviews with Philip Armstrong, Isabel Conklin, Sears Crowell, Robert Kahler, Donald Lancefield, Paul Reznikoff, George T. Scott and Elsie Scott, Horace Stunkard, Albert Szent-Györgyi, and William Randolf Taylor proved very valuable in filling out details and putting much of what we had learned in the dry, archival form of official records into personal terms.

In addition, Betsy Bang, Bob Barlow, Sears Crowell, James Ebert, Jane Fessenden, Donald Lahy, Dorothy Rogers, Homer Smith, Susie Steinbach, J. P. Trinkaus, and John Valois helped fill in details about times and attitudes during their respective careers here. Undoubtedly, others would willingly have helped if this had been a different sort of project and if they had been asked. Again, as a historian, I encourage everyone who has some further or alternative piece of the story of the MBL to record that information and deposit it in the Archives for future researchers.

Aside from the specific references given below, quotations or other information often came from the official records of the MBL. Frank Lillie's *The Woods Hole Marine Biological Laboratory*, published in 1944 and reprinted by the *Biological Bulletin* in 1988 (v. 174), is extremely valuable for historical detail. I have not reiterated the many facts and chronologies included there.

In addition, the annual reports and the trustees' minutes contain invaluable information, though seldom as much of the story behind the official, sanitized conclusions as one would like. The Lillie Papers, given to the MBL Archives by the University of Chicago, contain his correspondence in preparing his history and help to fill out some of the discussion between the official lines. Other collections of letters and assorted documents give additional perspective. The *Collecting Net* provides insight into the less formal, human side of MBL events.

The official records contain myriads of wonderful data: who came, what schools and what countries they represented, for what purposes, how many women, what research they published, what lectures were given, and so forth. There are even compilations of statistics for the early decades. Quantitative historians should take note of this largely untapped resource.

Many photographs included here also come from the MBL Archives. Most have resided there for some time, the product of past professional and enthusiastic amateur photographers. A few have been given recently, in anticipation of centennial projects. It is hoped that other people will consider cleaning out the attic and adding to the extremely impressive collection of many thousands of photos. Each photo collection given recently has been kept intact by donor or photographer to maintain the integrity of the collection. For earlier materials, given over many years and sometimes with no official records kept, every effort has been made to identify photographers and donors. MBL archivist Ruth Davis has done all the hard work of sorting through the thousands of photos, selecting the most appropriate, identifying them, and putting them in order.

The earliest formal photographs were taken by professional Boston photographer Baldwin Coolidge. The MBL evidently hired him to take representative but interesting formal group pictures beginning in the very first years. Many of the glass plate negatives of this first-rate photographer's work are now held by and provided to us courtesy of the Society for the Preservation of New England Antiquities. An apprentice of Coolidge's, Howard S. Brode, who was also a student at the MBL, followed him around and took similar shots. Brode's photos make an informative contrast, as they were never quite as effective. For further discussion of Coolidge, see Jane A. McLaughlin, "Baldwin Coolidge, Photographer 1845–1928," *Spritsail, Journal of Falmouth History* (1987) Vol. 1 No. 1:0 5–25.

Then comes the highly informal collection of Gideon Dodds, given in the form of a photo scrapbook, which documents life at the MBL in the 1920s. Other similar informal photo albums provide other individual pictures. In addition, Isabel Conklin recently donated a set of photos depicting life around Woods Hole, including pictures of her father, who played such an important role in the early life here.

The Public Information Department, over the years, has accumulated a set of pictures. Unfortunately, many were taken by unnamed staff members in connection with one or another publicity project then underway. A series for *Life* magazine, for example, or some for a WGBH program or a special ceremony or exhibit: photos taken for many different reasons and by many different people have found their way into the Archives, and some of those are in this book.

In working on this project, we added a few new discoveries as well. Undoubtedly there are others to be made. Elaine Pear Cohen provided the photograph of her bonded bronze 1980 sculpture, "Woods Hole: The Scientists." The sculpture resides in the collection of Dr. Virginia Peters, on the corner of Water and School streets in Woods Hole and was photographed by Sally Brucker. Deborah Day, archivist at Scripps Institution for Oceanography, sent the photograph of Purdum as Gosnold.

Perhaps the most exciting find, because of its quality and extent, is the work of Alfred Francis Huettner. An accidental meeting with MBL Associates Robert and Millie Huettner, now of Woods Hole, revealed that Robert's father had, in fact, taken many of the beautifully artistic photos of people and activities at the MBL during the early decades of this century. Many people have long wondered who had made such excellent portraits as those of E. B. Wilson and T. H. Morgan that reside in the Archives and have hung on MBL walls and have been printed in biographical works on these men. Now we know. The Huettners have negatives as well as quality prints that beautifully illustrate the MBL life and people. They have generously shared those with the MBL, and modern prints and negatives are now also on deposit in the Archives. An exhibit of Huettner's work graced the Meigs Room walls during the centennial summer, and the prints are now scattered throughout the MBL. For more discussion of Alfred Francis Huettner, see *Collecting Net* (August 1987) 5#5: 4.

In working with the historical collection of photographs and preparing modern negatives and prints for archival purposes and for this volume, Linda Golder and Linda McCausland have provided invaluable help. Linda Golder heads the photolab at the MBL and has been very helpful in providing quality copies of older materials from the MBL collection. Linda McCausland, a professional photographer in Orleans, Massachusetts, has worked with the old and fragile materials requiring special care. With her previous experience at the Eastman School of Photography working with archival photos, she has provided the expertise to prepare negatives and prints of far better quality than the original, sometimes faded and yellowed, prints we have had in the archives. The careful work of both these highly qualified women has really made the photographic side of this project possible.

Thanks also to everyone else who has helped, perhaps without even realizing it: to those who have contributed materials; to those who have come in and been asked abruptly "who was this" or "where was that"; to those who have cheered for the project (especially John Pfeiffer, who egged Ruth and me on at discouraging times; the watchmen who offered support early on Sunday mornings or late at night; Jane Fessenden, who would not let us quit; and Joel Davis, who provided much needed sustenance at critical times); to those with good editorial advice, especially Phil Pauly, Richard Creath, and Garland Allen; to Public Information director George Liles, who provided much information and editorial assistance; and to centennial coordinator Pam Clapp, who really single-handedly kept things going at crucial points. I especially appreciate the support and coordination of Jones and Bartlett's editor, Joe Burns, for steering the book through to publication. Special thanks go to publisher Don Jones for his support and overall direction. I would also like to acknowledge Jones and Bartlett's production director, Maureen Cunningham Neumann, and her able staff, Rafael Millán and Anne Benaquist, for their skill and expertise in designing and creating this book. Most of all, thanks to everyone who has been part of the life of the MBL and who has therefore contributed in some way to this autobiography. This book has really been a community project in the spirit of the MBL.

Epilogue

In the early 1970s the American Institute of Biological Sciences published a collection of essays to mark the twenty-fifth anniversary of AIBS. I served on the editorial committee of that volume, which included twenty-one chapters written by leading scientists from nearly every area of biology. We asked our eminent authors not to look back over AIBS's first twenty-five years, but to look forward and speculate about the problems biologists would be tackling in the next twenty-five years. The resulting essays were provocative, ambitious, and, for the most part, wrong: nearly everything we hoped to accomplish in twenty-five years was accomplished within a decade.

Although it is always difficult to look forward in science with any precision, an anniversary—especially a centennial anniversary—is an appropriate time both for looking to the future and for remembering the past.

At the Marine Biological Laboratory, recalling the past has occupied numerous historians, philosophers, scientists, and writers over the last few years. Among those who have undertaken historical projects, Jane Maienschein and Ruth Davis have produced one of the most accessible and personal portraits of the MBL. Professor Maienschein's text and archivist Davis's pictures recall the men and women of science who founded the MBL, and the subsequent generations of investigators and students who came to work in the village that served as the crossroads of American biology. These, of course, are the stories you'd expect to find in a popular history of a science institution. Less predictably, *100 Years Exploring Life* remembers the nonscientific side of life at the early MBL: conversations in the old mess hall, songs sung by students at the MBL Club, canoe races, and amateur whaling expeditions. The book remembers people like Charles R. Crane, who rallied to the support of the fledgling laboratory in its early decades.

As you read about the sense of adventure and discovery that prevailed in the laboratory's early days, you begin to understand why Woods Hole residents who were not themselves directly engaged in science nonetheless welcomed the MBL—donating scarce parcels of land, erecting new buildings, lobbying with foundations on behalf of the sometimes financially pressed laboratory, and, on many occasions, writing checks to cover year-end budget shortfalls. In the early part of the century, the local community—everyone from merchants to boardinghouse owners, from fishermen to captains of American industry—understood that the MBL brought to the village a vitality and an intellectual excitement that wasn't to be found in ordinary fishing villages and oceanside resorts.

So one part of our centennial celebration has involved a long, lingering, often affectionate look back to see whence we have come. An equally important part of the centennial has been our effort to look forward, to ask what role the MBL will play in the next decades of American biology. We are entering our second century concurrent with the advent of the Age of Molecular Biology. The galloping progress that overtook the 1971 AIBS predictions has, if anything, picked up its pace, and today biologists are developing new tools and new applications at an unprecedented rate. Already, the powerful new tools of molecular biology have brought vastly improved techniques for fighting and preventing disease, refinements in fertility and population control, and the genetic engineering that has made possible the most profound developments in agriculture since humankind first domesticated plants and animals many thousand years ago. The revolutionary progress in our understanding of life on the most basic and useful levels will surely continue through the end of this century.

Fortunately, our task in looking forward is not to set a timetable for specific discoveries, but to clarify the role the MBL will play in modern science—an enterprise several orders of magnitude larger and vastly more complex than it was in C. O. Whitman's time. We must build the twenty-first-century MBL without losing sight of the nineteenth- and twentieth-century MBL. We must keep alive the old MBL—the informal, cooperative, adventuresome, free-spirited laboratory so affectionately described in the pages and photographs of this book.

The centennial look forward has involved a broad cross section of the community and expert consultants, including summer investigators and year-round scientists, neurobiologists and cell biologists and ecosystems analysts, our colleagues in the Boston University Marine Program, historians, friends from business and industry, national science policy leaders, community leaders from Woods Hole and Falmouth, and our friends in state government. Because we have had input from so many people who harbor so much affection and concern for the MBL, we can say with some confidence where the MBL is headed.

Clearly, our world-famous summer programs of teaching and research will remain our raison d'être. We will continue to modernize our facilities and to introduce new biological approaches in our summer courses and in our summer research programs.

We will continue to provide tutorial laboratory courses at the cutting edge of science. Given that the faculty and students who populate these courses are among the leaders in their fields throughout the world, it is unlikely that the same courses could be given at any one university. The MBL community is committed to the continuation of these one-of-a-kind courses, where the next generation of biologists is trained in the use of modern biological techniques and the value of marine organisms. To make this possible, we will seek over the next decade to raise endowment funds to cover course expenses that are not covered by current tuition or grants from foundations and the government.

We intend to enlarge the year-round research program and provide additional up-to-date laboratory space to accommodate the increased year-round staff. The expanded year-round programs will provide a stable base for summer programs of research and education. We anticipate developing a year-round critical mass in neurobiology, cell biology, and developmental biology, all with a common theme of molecular evolution. These areas will complement our already strong and still expanding year-round Ecosystems Center. The expanded year-round programs will share the modern approaches of molecular biology and molecular genetics.

The development of genetic engineering has brought a scientific revolution that is dominating the last quarter of our century in much the way physics dominated the first quarter. Genetic engineering has made possible new kinds of biological experiments that yield precise information about the basic mechanisms of life. Our new ability to compile libraries of genetic information has made it possible to compare organisms and to explore the evolution of life on a level that is providing new understandings of health and disease.

The powerful new tools of molecular biology can be used to work on a wide range of biological problems—from general questions of phylogeny to very specific questions, such as of the origins of neural receptor sites. Recognizing the power and range of molecular biology, we intend to develop, in cooperation with our sister institution, the Woods Hole Oceanographic Institution, a center for molecular evolution that will serve as the basic platform on which to integrate the disciplines of neurobiology, cell biology, and developmental biology.

Along with the exciting across-the-board advances in late twentieth-century biology, another recent development holds great promise for the MBL: the growing recognition that many biological disciplines can be studied through the use of marine models, which offer good views into basic life processes and reduce the need for use of warm-blooded animals in research. Easy to study, inexpensive, elegantly simple and remarkably diverse, marine plants and animals are attracting increasing interest from federal agencies and private foundations. With its long history of educational and research programs and its continuing reputation as the nation's premier marine laboratory, the MBL will remain in the vanguard of biological research.

We plan to develop facilities and expertise for culturing, rearing, and studying the genetics of those marine animals that are so valuable as biomedical models. At the same time, to complement modern molecular approaches, we will need an updated and modernized facility for the warm-blooded animals required for the preparation of monoclonal antibodies.

This merging of basic research, biomedicine, and marine biology builds upon the traditions celebrated in Jane Maienschein's text and the photographs selected by Ruth Davis. Readers of *100 Years Exploring Life* will note that our plan for the next few decades is not a new venture, but a reaffirmation of the MBL's mission and a focusing of its resources.

Of course, while we are expanding programs and modernizing research facilities, we will continue to nourish our traditional programs and resources. We will maintain our relationship with the Boston University Marine Program, a remarkably symbiotic arrangement that for nearly two decades has added to traditional MBL strengths in cell biology and neurobiology, provided us with another window on environmental science, and given us a continuous tie to academia.

Professor Maienschein's text devotes an entire chapter to the MBL Library, a facility beloved by several generations of scientists and historians. We will continue to operate this unparalleled marine and biological science library, which is jointly supported by the several institutions within the Woods Hole scientific community. Maintaining an up-to-date library is a major challenge. Much as biological research has been profoundly affected by the techniques of molecular biology, library and information science has been changed in our era by the burgeoning mass of published data and by computerized information storage and retrieval techniques. Several private foundations are working with us to update our library facilities and to network our library with major libraries across the country.

In our centennial year, we have paused in our work to acknowledge the contributions made by generations of MBL students, faculty, and investigators. But while we have honored our past, we have kept a focus on the future—looking forward and backward at the same time. This dual process of remembering whence we came and deciding where we are bound will continue beyond the centennial year, as it should in an institution as venerable and vibrant as the MBL.

HARLYN O. HALVORSON

Woods Hole
July 29, 1988